TUTORIALS ON MOTION PERCEPTION

NATO CONFERENCE SERIES

I   Ecology
II Systems Science
III Human Factors
IV Marine Sciences
V   Air-Sea Interactions
VI Materials Science

III HUMAN FACTORS

TUTORIALS ON MOTION PERCEPTION

Edited by
Alexander H. Wertheim
and
Willem A. Wagenaar
Institute for Perception TNO
Soesterberg, The Netherlands

and
Herschel W. Leibowitz
Pennsylvania State University
University Park, Pennsylvania

Published in cooperation with NATO Scientific Affairs Division

PLENUM PRESS · NEW YORK AND LONDON

Library of Congress Cataloging in Publication Data

Main entry under title:

Tutorials on motion perception.
     (NATO conference series. III, Human factors; v. 20)
     "Derived from the NATO symposium entitled 'Symposium on the Study of Motion
Perception: Recent Developments and Applications,' held August 24–29, 1980, in
Veldhoven, The Netherlands''—T.p. verso.
     Symposium sponsored by NATO.
     Includes bibliographical references and index.
     1. Motion perception (Vision)—Congresses. I. Wertheim, Alexander H. II.
Wagenaar, Willem Albert, 1941 –   . III. Leibowitz, Herschel W. IV. Symposium
on the study of Motion Perception: Recent Developments and Applications (1980:
Veldhoven, Netherlands). V. North Atlantic Treaty Organization. VI. Series.
BF241.T87 1982   152.1'425   82-16554
ISBN-13:978-1-4613-3571-9   e-ISBN-13:978-1-4613-3569-6

DOI:10.1007/978-1-4613-3569-6

Derived from the NATO symposium entitled ''Symposium on the Study
of Motion Perception: Recent Developments and Applications,''
held August 24–29, 1980, in Veldhoven, The Netherlands

©1982 Plenum Press, New York
Softcover reprint of the hardcover 1st edition 1982

A Division of Plenum Publishing Corporation
233 Spring Street, New York, N.Y. 10013

PREFACE

From August 24-29, 1980 the international "Symposium on the Study of Motion Perception; Recent Developments and Applications", sponsored by NATO and organized by the editors of this book, was held in Veldhoven, the Netherlands.

The meeting was attended by about eighty scholars, including psychologists, neurologists, physicists and other scientists, from fourteen different countries. During the symposium some fifty research papers were presented and a series of tutorial review papers were read and discussed. The research presentations have been published in a special issue of the international journal of psychonomics "Acta Psychologica" (Vol. 48, 1981). The present book is a compilation of the tutorial papers.

The tutorials were arranged around early versions of the chapters now appearing in this book. The long discussions at the Veldhoven tutorial sessions resulted in extensive revisions of the texts prior to this publication. Unfortunately this led to a delay in publication, but we feel that this was justified by a greater depth of understanding which, in our opinion, has significantly increased the quality of the book. As they now stand, the chapters cover most of the issues relevant to the study of motion perception. Also they clearly reflect the intensive exchange of knowledge that took place during the symposium. As such we think that this book can be used both as an advanced text for students and scientists alike and as a comprehensive reference source.

The chapters of this book have been ordered in a way that reflects to a certain extent the history of interest in the subject of motion perception.

The first two chapters are written by two of the foremost pioneers and veterans of the field, Hans Wallach and Gunnar Johansson. Each of them has established what can be recognized as a whole tradition of research. Both authors show that these traditions are still very much alive and have the potential of continuously broadening the scope of the psychological approach to motion perception.

In the third chapter Claude Bonnet adds to these theories by reviewing the experimental methodology needed for the actual measurement of perceived motion. In addition, his review brings together an enormous amount of experimental data, considered in terms of a theoretical framework characterized by two systems responsible for movement perception, a framework thoroughly discussed in Veldhoven.

The following chapter, by Robert Sekuler and his co-workers, is close to these issues. It reviews recent experimental work and theoretical issues related to the complex interactions between the perception of motion and of direction.

From here it is but a small step to Len Matin's contribution on the subject of perceived stimulus location, with special emphasis on the somewhat controversial concept of extra-retinal eye position information. The review includes almost everything written on this subject, thereby filling a long-felt gap in the literature.

In the next chapter Alain Berthoz and Jaques Droulez address the issue of visual-vestibular interactions, with special emphasis on the role of the otolith system. All current models are examined against the available experimental evidence, much of which is insufficiently known to scientists working exclusively in the visual domain.

Mark Berkley's contribution, which is next, focusses on neurophysiology. It should be read as a review of experimental data, written with the purpose of establishing their relevance to the existing theories of motion perception.

The last chapter was written jointly by Herschel Leibowitz, Bob Post, Thomas Brandt and Johannes Dichgans. It includes the subject matter of several of the tutorial papers discussed in Veldhoven, such as visual-vestibular interactions in circular vection, simulators and human factors in vehicle guidance. Its main message is its argument against a distinction between "basic" and "applied" research, a conclusion which in our opinion cannot be overstressed, and which faithfully summarizes the main thrust of the Veldhoven meeting.

We hope that the reader will enjoy this book as much as we enjoyed discussing its topics during the symposium.

Alex H. Wertheim

Institute for Perception TNO

Soesterberg, The Netherlands

ACKNOWLEDGEMENTS

We would like to express our gratitude to NATO whose financial aid enabled both the Symposium and the publication of this book. We also wish to thank Margie Knopper from our institute, who spent many hours typing, retyping and correcting the various versions of the chapters of this book, most of which was done in her spare time. Without her persisting stamina the publication of this book would have been virtually impossible. We also thank Carol Varey who was of great help as our English language editor and Koos Wolff for his skilled work in adapting some of the graphic contributions. Many others have been helpful in the making of this book. Among them we want to mention dr. W. Bles of the Free University of Amsterdam, Dr. G. Orban of the University of Leuven (Belgium) and Drs. P. Padmos and J.B.J. Riemersma of our own institute, for their helpful comments on several of the papers. It was a pleasure to work together with the authors themselves, who were always happy to comply with the demands made upon them by the editors and who sometimes helped us with editorial intricacies. In addition it should be mentioned that we feel indebted to the National Defence Research Organization TNO and particularly the Institute for Perception TNO which provided the necessary time and facilities to organize the Symposium and to edit the proceedings. Finally, we would like to express our gratitude to the authors who granted us permission to reproduce material from their papers and books and the following publishers for allowing us to use their copyright material:

Academie des Sciences, Paris (C.R.T. 278 - Séance du 18 Mars 1974, Serie D: *Role de la Vision Peripherique et interactions visuo-vestibulaire dans la perception du movement lineaire chez l'homme*).

Aerospace Medical Association, Washington D.C. *(Aviation Space and Environmental Medicine)*

Almqvist and Wiksal Periodical Comp., Stockholm *(Acta Otolaryngologica)*

The American Academy of Optometry, Baltimore *(Transactions of the American Academy of Optometry)*

The American Association for the Advancement of Science, Washington D.C. *(Science)*

The American Physiological Society, Bethesda *(Journal of Neurophysiology)*

The Association for research on vision and ophthalmology Inc, St. Louis (*Investigative Ophthalmology and Visual Science*)

Cambridge University Press, Cambridge (*Applications of Psychophysics to Clinical Problems; Journal of Physiology*)

Columbia University, New York (Ph.D. diss. J. Pola: *The relation of visual Direction to Eye Position during and following a Voluntary Saccade*)

Lawrence Erlbaum Ass., Hillsdale (R.A. Monty and J.W. Senders Editors: *Eye movements and Psychological processes*)

Junk Publishers B.V., The Hague (*Documenta Ophthalmologica*)

Alan R. Liss, Inc., New York (R. Sekuler, D. Kline and K. Dismukes Editors: *Aging in Human Visual Functions*)

MIT Cambridge, Mass. (Ph.D. Thesis C.C. Ormsby: *Model of Human Dynamic Orientation*; Masters Thesis W.H.N. Chu: *Dynamic Response of Human Linear Vection*)

Naval Aerospace Medical Institute, Pensacola (*Monograph, 14, 1966; NASA SP-152 (1967)*)

Pion Publications Ltd., London (*Perception*)

The Royal Society, London (*Proceedings of the Royal Society*)

Springer Verlag, Heidelberg (*Experimental Brain Research*) (D. Jameson and L. Hurvich, Editors: *Handbook of Sensory Physiology VII/4D*); (R. Hold, H.W. Leibowitz, and H. Teuber, Editors: *Handbook of Sensory Physiology VIII*); (E. Cool and E. Smit, Editors: *Frontiers in Visual Science*)

Verlag Schwappach, Ganting (*Fortschritte der Medizin*)

CONTENTS

NEURAL SUBSTRATES OF THE VISUAL PERCEPTION OF MOVEMENT   201
     Mark A. Berkley

CONTENTS

IMPLICATIONS OF RECENT DEVELOPMENTS IN DYNAMIC SPATIAL   231
ORIENTATION AND VISUAL RESOLUTION FOR VEHICLE GUIDANCE
Herschel W. Leibowitz, Robert B. Post,
Thomas Brandt and Johannes Dichgans

EYE MOVEMENT AND MOTION PERCEPTION*

Hans Wallach

Swarthmore College
Swarthmore, Pennsylvania 19081
United States of America

SUMMARY

Two procedures are described with which the relative effectiveness of the three conditions of stimulation that cause motion perception can be explored, and the results of such experiments are
reported.
    Experienced motions are described that result when ocular
pursuit and object-relative displacement are stimuli for different
components of the motion of an object and when perceived motion of
the object is therefore the outcome of different perceptual processes that are simultaneously in operation.
    An investigation of the nature of compensatory eye movements
that take place when the head is moved is reported.

1. THREE STIMULI FOR MOTION COMPARED

    Eye movements are only one of three conditions of stimulation
that cause motion perception and are, in different ways, less important than the other two. Only rarely do eye movements function
as the sole cause of motion perception. Most of the time, they
occur when other conditions of stimulation for motion perception
also operate. Therefore, this discussion starts with motion perception in general.

---

* The experiments here reported were supported by Grant 11089 from
  the National Institute of Mental Health, USA to Swarthmore
  College, Hans Wallach, principal investigator.

Duncker (1929) recognized that there are two conditions that lead to the perception of motion, the displacement of the moving object relative to stationary objects and its displacement relative to the observer. The two conditions give rise to three conditions of stimulation. Duncker's object-relative displacement makes itself felt in a changing configuration on the retina, while there are two conditions that can represent the subject-relative displacement of the moving object. One is displacement of the object's image on the retinae when the eyes fixate a stationary object, and the other consists in pursuit eye movement that occurs when the eyes fasten on the moving object.

It seems that there can be no question about the effectiveness of ocular pursuit as a stimulus condition for perceived motion. When the moving object is given in a homogeneous environment, the eyes rapidly take up tracking the moving object, and, from that point on, pursuit is the only manner in which the object's motion can be effectively perceived. That both image displacement and configurational change are stimuli for motion perception is less obvious. Either one would suffice. When the eyes do not track the moving object and rest instead on a stationary object that is present, the image of the moving object is displaced on the retina. Hence the two subject-relative conditions of stimulation, pursuit and image displacement, together can account for nearly all veridical motion perception.

There is, however, evidence that the object-relative condition of stimulation, configurational change, is a strong factor in motion perception. A stationary object that is surrounded by an extended pattern that is in motion is usually seen to move in the direction opposite to the motion of the surround. While this so-called induced motion is seen, the object is represented as stationary by the subject-relative conditions: when the eyes rest on the stationary object, the absence of eye movements indicates that the object is actually at rest. Nevertheless, induced motion is seen in the majority of cases. Object-relative and subject-relative stimulation are here in conflict, and the former is usually winning out.

There is still another fact that shows the importance of configurational change in motion perception. Under appropriate conditions, perceived speed depends to a much greater degree on the rate of the configurational change by which an object's motion is given than on the rate of the moving object's subject-relative displacement. This fact emerges from a discovery by J.F. Brown (1931). Brown had subjects match the motion speeds of spots that moved through illuminated apertures of different sizes in a darkened room. The speed matches resulted in velocities of the objective motions that were nearly proportional to the sizes of the apertures. (This "transposition effect" has often been viewed as an artifact. It was supposed to result from rows of evenly spaced spots moving through the aperture and leaving it at regular time intervals. The speed impression set up by this rhythmic event was held responsible for the transposition effect, e.g. Kling and Riggs, 1971, p. 521. Proponents of this view overlooked that Brown repeated all experiments with only single spots visible, with the expressed purpose to obviate this interpretation.) Brown found that if subjects compared speeds in apertures of different sizes where the sizes of the moving spots remained in the same ratio to the aperture size, then speed came close to being the same when the objective motion velocities were proportional to these sizes. In other words, when "motion fields" were of identical shapes and differed only in size, spots that took equal time to traverse their apertures seemed to move with nearly equal speeds. Motions that caused the same changes per unit of time of the configurations formed by the spots and their apertures appeared to have almost the same speeds. To a great extent, speed depended on the rate of configurational change, and only in a minor way did it depend on the rate of subject-relative displacement. As in the induced motion paradigm, two conditions of stimulation conflict with each other in Brown's experiment. To the extent that perceived speed depends on subject-relative displacement, perceived speeds should be equal when the absolute velocities are the same, and to the extent that configurational change prevails, perceived speeds should be equal when the rates of change of configuration, regardless of size, are the same. An inspection of the speed matches for different size ratios obtained by Brown leads to the rough estimate that configurational change contributed 80%, and subject-relative displacement 20%, to the perception of motion speed.

By its nature configurational change is relative; the displacement between an object and its surround can result from motion of the object or from motion of the surround, and either one of these motions can be perceived. Why then, is the stationary object in induced motion seen to move? Duncker formulated a rule about the manner in which configurational change brings about perceived motion: When relative displacement between an object and its surround is given, motion will be assigned to the object and immobility to the surround. This assignment of motion and rest makes configurational change a veridical stimulus condition in ordinary motion perception, namely, when a moving object is displaced relative to the stationary environment. Being given in relative displacement to a surround, the moving object is seen to move.

Due to Duncker's rule, configurational change suffices to bring about veridical perception of motion, when the moving object is visible among stationary objects. The subject-relative stimulus conditions are unnecessary. In fact, motion too slow to be perceived by ocular pursuit may be perceived due to configurational change (Mack, Fisher & Fendrick, 1975). Of the two subject-relative stimulus conditions only one, ocular pursuit, is indispensable; as stated above, it causes motion perception when stationary objects are absent. Whether image displacement actually plays a role in motion perception is a harder question to answer, for it cannot be studied in isolation. It requires the presence of a stationary object that can be fixated; the presence of a stationary object introduces configurational change and therefore makes an interpretation difficult.

Under ordinary viewing conditions when the moving object is seen in an array of stationary objects or against a patterned background, its perceived motion is predominantly determined by configurational change. This can be inferred from experiments with induced motion where motion is seen according to configurational change in spite of contradicting subject-relative conditions. When a stationary object is seen against an extended pattern in translatory motion, induced motion is seen by more than 90 percent of the subjects. There are even conditions where configurational change always prevails. That is the case when the object that is surrounded by a moving pattern is given an objective motion of its own. If, for instance, an extended pattern moves horizontally and a dot within the pattern simultaneously moves vertically, the dot will then be obliquely displaced in relation to the pattern. Under these conditions the dot always appears to move obliquely, in the direction of its displacement relative to the pattern. The apparent oblique motion of the dot is seen even when the subject tracks its objective vertical motion. The configurational change, that is, the oblique displacement of the dot relative to the pattern, always prevails over the vertical pursuit with which the dot's motion is given subject-relatively.

A more interesting result is obtained when the moving pattern, in relation to which the vertically moving dot undergoes an oblique object-relative displacement, is so altered that the objective vertical motion of the dot is not given object-relatively at all. That can be done by changing the moving pattern in such a way that it offers no landmarks for the vertical motion of the dot, namely, by using a pattern that consists of vertical lines only. Now, when the eyes track the vertically moving dot, its motion is given solely by ocular pursuit. At the same time, the horizontal motion of the vertical lines of the pattern causes a horizontal displacement between the dots and the lines. This relative displacement has peculiar properties. It does not define a motion of the dot in a specific direction. Since the pattern of vertical lines offers no landmarks for motion in vertical direction, the displacement between the dot and the lines is defined only as to dimension and as to sense; the dimension is horizontal and the sense is opposite to that of the motion of the line pattern. In other words, the displacement between the dot and the line pattern defines a variety of directions that have a horizontal component opposite to the motion of the pattern. All of these motion directions are equally justified where the displacement between the dot and the line pattern is concerned. Which of these potential directions can be seen depends on what is subject-relatively given. When, as is here the case, the dot is subject-relatively in vertical displacement, any oblique motion of the dot is compatible with the given vertical displacement if the oblique motion has a vertical component that is equal to the displacement. The horizontal component of the oblique motion, on the other hand, depends on the effectiveness of the relative displacement between the dot and the line pattern.

Wallach, Bacon and Schulman (1978) presented subjects with such an arrangement and found that the dot was seen to move on a fairly smooth oblique path. Subjects were able to give estimates of the tilt of this apparent motion path, by adjusting the tilt of a test rod that was attached perpendicularly to a horizontal shaft and could be given any desired orientation in the subject's frontal plane. When both screen and dot underwent the same reciprocating motion, the dot moving up and down 10 times per minute over a distance of 15 cm and the line pattern moving left and right over the same distance in phase with the dot's motion, the mean tilt estimate was 43.9°(N=25). This was very near the direction of the motion path of 45° that would be expected if configurational change is fully effective, that is, if it caused just as great a horizontal motion of the dot as the vertical motion that resulted from the pursuit of its vertical excursion. A motion of 45° tilt would, of course, not normally result when a vertical motion is subject-relatively given. As is always the case when conditions for induced motion are given, a conflict exists between subject-relative stimulation, which represents the dot's motion as vertical and object-relative displacement which, if fully effective, would produce motion with a 45 degree tilt. The fact that the mean tilt estimate was close to 45° shows that object-relative displacement prevailed. As in the previously described case of induced motion, the conflict between configurational change and ocular pursuit was resolved in favor of configurational change.

Does the other stimulus condition that represents subject-relative displacement, namely, the displacement of the moving object's retinal image, yield just as readily to configurational change? We added a small stationary mark to the arrangement that was just described and had the subject fixate it. This caused the vertical motion of the dot, which previously was tracked, to be given as an image displacement. The stationary mark was projected on the moving line pattern 2 cm to the side of the dots vertical motion path and 2.5 cm above its center. As in the previous experiments, the vertical excursion of the dot and the horizontal excursion of the vertical line pattern were equal. Changing from pursuit of the vertical dot motion to having it given as image displacement had a striking effect. No longer was the conflict between object-relative and subject-relative motion perception, which is inherent in induced motion, resolved in favor of the former. When the vertical dot motion was tracked, the mean tilt estimate had been 43.9°; with the motion given as image displacement it was 18.9°(N=16). This mean tilt estimate was still significantly different from 0°($t(15) = 8.76$), but it was much different from 43.9°. Whereas the domination of object-relative motion perception over ocular pursuit was virtually complete, now the shift of the perceived motion direction away from verticality amounted only to 34.2 percent of the shift that full domination by configurational change would have produced.

As a control, the same 16 subjects who had given tilt estimates when they fixated the stationary mark also gave estimates
while they tracked the vertically moving dot, but with the stationary mark present. Under these conditions the mean tilt estimate
was 27.7°. This result was significantly different from 18.9°, the
mean tilt estimate that had been obtained with fixation (t(15) =
4.72; p <.001) and shows that the effect of fixation was not merely
due to the presence of the stationary mark. But the fact that the
mere presence of the stationary mark near the path of the tracked
dot changed the latter's mean motion direction from 43.9° to 27.7°
is remarkable and currently under investigation.

Two facts emerged from these simple experiments: (1) In conflict with configurational change, image displacement was a more
potent cue for subject-relative displacement than ocular pursuit.
(2) When vertical pursuit and horizontal object-relative displacement were motion stimuli for the same object and two different
motion processes came simultaneously into operation, their results
combined to produce a fairly smooth and complete path. We have
been interested in this higher order motion perception.

2. PATHS THAT RESULT FROM A COMBINATION OF INDEPENDENT MOTION PROCESSES

The two objective reciprocating motions that were presented in
the experiment by Wallach, Bacon and Schulman, the vertical motion
of the dot and the horizontal motion of the line pattern, were
simple harmonic. Therefore, the oblique motion path that was here
perceived would have been a simple Lissajous figure, had it been
objectively given. But since it emerged as the combination of the
results of two independent perceptual processes, the question arose
as to how accurate such perceptual Lissajous figures are. If one causes a
dot to reciprocate in simple harmonic motions simultaneously in two
directions that are perpendicular to each other, the resulting objective motion paths can have a variety of elliptic shapes, including an oblique straight line and a circle. Which shape results
depends on the relative extents of the component motions and their
phase relations. More complex shapes result, when one changes the
frequency ratio of the two component motions from 1:1 to one where
one component has twice the frequency of the other (2:1) or if one
uses a frequency ratio of 3:2. We found that the same variety of
perceived motion paths are perceived when the combination of the
two motion components is psychological and results from two independent perceptual processes.

We did an experiment in which the frequencies of a
vertical and of a horizontal motion component were the same,
but where the horizontal component started each excursion when the
other component motion was at its midpoint. In real Lissajous
figures of this sort, that is, when a dot actually undergoes the
two component motions simultaneously, elliptical paths result. The

shapes of the paths depend on the relative extents of the two component motions. When the horizontal excursion is larger, the result
is an elliptical path with the longer axis horizontal, and when it
is shorter, the elliptical path is in vertical orientation. When the
two component motions are equal, the path is circular. In our experiment, this combination of two motions was psychological. A dot moved
vertically and its motion was given by ocular pursuit; the vertical
line patterns on which it was visible moved horizontally and caused
induced motion of the dot in the horizontal dimension. The line pattern always started its excursion when the dot was at the midpoint
of its motion path. The excursion of the line pattern could be made
greater or smaller. The primary purpose of the experiment was to
determine the ratio of the extent of the horizontal motion of the
line pattern to the extent of the vertical motion of the dot at
which a subject reported a circular path. That ratio depends on
the relative effectiveness of the motion processes that result
from ocular pursuit and from configurational change and represents
a measurement of the relative effectiveness of the two stimulus
conditions. Here, we are not concerned with these ratios. Rather,
we are interested in the accuracy with which individual subjects
distinguished a circular path from an elliptic one, for we are
concerned with the consistency with which the different perceptual
processes combine to produce a single motion path. A set of different extents of the horizontal motion of the line pattern was
repeatedly presented to each subject in an abbreviated method of
constant stimuli, and a range over which circle judgments occurred
was determined. There were large individual differences. Expressed
on a scale of ratios of vertical and horizontal motion extents,
the range over which circle judgments occurred varied from cases
where only a single ratio elicited a circle judgment to a range
of .16 extent ratios wide. The average range of extent ratios in
which circle judgments occurred was .08 (N=37).

Just how accurate perceptual Lissajous figures of frequency
ratios other than 1:1 are perceived will be known only when their
perception can be compared with the perception of corresponding
real Lissajous figures, for we do not know how well the latter are
apprehended. We plan to obtain drawings and descriptions of the
motion paths that are perceived when dots move on real Lissajous
paths and compare them with drawings and descriptions of the
paths of analogous perceptual Lissajous figures.

3. COMPARISON OF THREE STIMULI FOR MOTION CONTINUED

When we wanted to compare the effectiveness of the three
stimuli that mediate motion perception, we experimented with induced
motion, because in induced motion the subject-relative conditions
of stimulation are in conflict with configurational change. Induced
motion is, however, not the only instance of such a conflict.
Gunnar Johansson (1950) discovered a number of motion arrangements

where this is also the case, and which lend themselves to comparisons similar to those that were made with our induced motion
arrangement. Johansson's arrangements differ from the induced motion
condition we used in that configurational change is not as potent
as in motion induced by extended patterns.

Johansson discovered a number of arrangements of moving spots
where the motions that are perceived result from a vector analysis
of the objective motion paths so that each objective motion is represented by two perceived motions. In the simplest of these arrangements, two spots move back and forth simultaneously and in
phase with each other, one vertically and the other horizontally.
One moves, for instance, downwards when the other moves to the left,
and they reverse their motions simultaneously. They briefly coincide
when their paths cross in the centers of each excursion. These crossing paths are, however, not perceived when the arrangement is freely
viewed; instead two component motions are experienced. The spots seem
to move straight towards each other, in our example one moves to SE
and the other to NW on the same path. They meet and cross, still
travelling in their original directions. At the same time, both
spots may be seen to move from NE to SW, that is, at right
angles to the path on which they seem to move toward each other.
In the case of each spot, the kinematic combination of the two perceived paths tallies with the vertical or horizontal path that is
actually given.

For a long time this splitting up into two component motions
was regarded as an ultimate fact. An explanation that I once proposed (Wallach, 1976) and which has been adopted by Proffitt,
Cutting and Stier (1979) is unsatisfactory. The following explanation makes more sense. It assumes that the spots that make up the
moving pattern form a group prior to motion perception, and that
the displacement of the group as a whole, which is subject-
relatively given, gives rise to one of the perceived motions. The
other perceived motion results from the relative displacements
within the group, which are given as configurational change. In
the present example the two spots as a group are objectively displaced between NE and SW. That displacement gives rise to one of
the perceived motions. The displacement of the spots relative to
each other is on an oblique line connecting them. It is given as
configurational change and gives rise to the other perceived
motions. The displacement of the group as a whole, on the other
hand, is only subject-relatively given. Thus, different kinds of
motion processes give rise to the component motions that are perceived. Thus, vector analysis takes place and different motion
vectors are perceived because different kinds of processes cause
them; the two component processes come into play for different
reasons.

The cue conflict occurs here in the following fashion. To be
sure, the motion of the group as a whole is given only subject-
relatively, and only one kind of motion process is possible. The
motions of the individual spots, however, are given not only

object-relatively but also subject-relatively, one as vertical and the other as horizontal. The fact that perceived motion is here in accordance with object-relative displacement is due to the potency of the process that is based on configurational change. Thus, induced motion and the kind of motion arrangements in which Johansson's vector analysis takes place have this in common: the processes based on configurational change cause perceived motions that differ from the motions caused by the processes that result from subject-relative conditions of stimulation. In the case of induced motion, an object that is subject-relatively given as stationary is seen to move due to configurational change. In the motion arrangement where the motion paths form a cross, configurational change causes the two points to move on the same path, toward and past each other, while subject-relative stimuli would cause them to move in different directions. Such differences do not occur in ordinary motion perception where the results of object-relative and of subject-relative stimulation are in agreement. Only in induced motion and in Johansson's arrangements will configurational change produce perceived motion that is different from that produced by the subject-relative stimulus conditions.

The cross arrangement was observed under three conditions. Either subjects viewed the motions freely, or one of the two moving points was tracked, or a stationary point was fixated and the objective motion paths were given as image displacements on the retinae. There were also two control conditions in which the stationary point was present but was not fixated, one for the free viewing condition and the other for the tracking condition. The subject described the motion patterns he saw and estimated the angles of the perceived motion paths.

The moving points consisted of small light spots projected on a translucent screen from the rear. They were 2.5 mm in diameter, the length of each motion path was 5 cm, and the display subtended a visual angle of 2.4^{o}. Motions were simple harmonic and each full cycle lasted 6 seconds. In addition to the two moving spots, a smaller and dimmer stationary spot could be projected on the screen. If visible, it was located 1.8 cm to the left of the upper end of the path of the vertically moving spot.

Two methods were used for obtaining estimates of the perceived motion directions. In one, the angle between the apparent motion paths was reproduced. A pair of flat metal bars joined at the centers could be made to form any desired angle. The subject arranged the bars so that they duplicated the angle between the perceived motion paths. In the other method subjects gave separate estimates of the slopes of the motion path of each spot by adjusting the tilt of a rod that could be given the desired orientation.

There were five viewing conditions that were always presented in the order listed.

    A. The moving spots were freely viewed.

    B. The stationary spot was also visible, but no fixation instructions were given.

C. Ocular pursuit of one of the moving spots was ordered.
    Alternate subjects tracked the vertical or the horizontal
    spot.
D. As condition C, but with the stationary spot present.
E. The stationary spot was present and fixated.
Eighteen subjects participated.

In condition A, where only the moving spots were visible and
the subjects observed freely, thirteen out of the 16 subjects who
gave consistent responses reported the spots to move straight toward
and past each other on a common path. The remaining three subjects
reported motion paths that intersected with sharp angles. Their
estimates of these angles which they gave by adjusting the crossing
bars amounted to 4.9^o, 6.8^o and 30.0^o. The first two of these
estimates can be regarded as showing complete vector analysis. Thus,
the reports of 15 out of 16 subjects conformed to Johansson's description of how one of the component motions is perceived. Seven
of these 15 subjects spontaneously reported the second component
motion, and this more or less agrees with Johansson's finding that
about 65% of the subjects report the second component.

Table 1. Means and standard deviations of the angles between the
          vertical and the tilt estimates of the motion paths of
          the tracked and of the un-tracked spot, in degrees.

Condition	N	tracked		un-tracked	
		Mean	SD	Mean	SD
C track vertical	9	15.2	10.1	58.0	16.2
track horizontal	9	74.6	14.3	41.6	7.2
D with stat. point					
track vertical	9	5.7	7.2	76.6	13.2
track horizontal	9	81.1	9.9	28.2	12.8
E stat. fixated	18	Mean	SD		
vertical		7.2	8.2		
horizontal		85.4	8.6		

The results of condition B could not be interpreted, because
we did not know how many subjects rested their eyes on the stationary point that was here added and ceased to view the display freely.

Having the subject follow one of the moving spots with his
eyes (condition C) brought about a strong change. Only one subject
reported that the two spots moved on the common path. In the case
of the 17 other subjects, tracking one of the spots caused its
motion to change somewhat toward its objective motion direction
which was either vertical (0°) or horizontal (90°). The means of

the tilt estimates for the tracked spots are listed in the first column of Table 1 and show that the perceived motion paths of the tracked spots deviated from 0° or from 90° on the average by 15.2° and by 15.4°. Perceiving the spots move according to configurational change, that is, on the common path, would have manifested itself in tilt estimates of approximately 45°. (This was approximately true of the mean tilt estimates for the untracked spot that are listed in the third column of Table 1.)

Thus, pursuit of a moving spot strongly changed the direction of its motion path. This effect of tracking was radically different from the result obtained by Wallach, Bacon and Schulman where induced motion, that is, configurational change, remained fully effective when the moving dot was tracked. That the mean tilt estimates for the tracked spot also differed significantly (p <.001) from the objective directions of the motion paths, that is, from 0° and 90°, shows that configurational change between the spots was also effective.

In condition E, where subjects fixated a stationary mark, both objective motion paths were given as image displacements on the retinae. The mean tilt estimates, listed in the last row of Table 1, were quite close to the objective motion directions of 0° and 90°. The configurational change between the spots had only a small effect, which was, however significant, with $t(17) = 3.42$ and p <.01. The mean tilt estimates were 7.2° and 85.4° instead of 0° and 90°.

Condition D was designed as a control for the presence of the stationary spot in condition E; the stationary spot was added to the two moving spots but was not fixated; instead, one of the moving spots was tracked. Adding the stationary spot had a strong effect. The mean tilt estimates for the paths of the tracked spots, as listed in Table 1, were here quite different from what they were in condition C. They were changed toward the vertical or the horizontal by 9.5° and 6.5° respectively. This change was so great that there was now no difference between the mean tilt estimates that were obtained for the tracked spots and those gotten when the motion paths were given as image displacements. Therefore we could not prove that image displacement was here more effective than pursuit eye movements in overcoming the effect of configurational change, although a comparison of the results of condition E with those of condition C shows this to be the case.

There was one striking difference between the results of the cross experiment and those reported earlier for the induced motion experiment. While configurational change was dominant over ocular pursuit in the induced motion condition, its effect was incomplete in the cross experiment. Here, tracking of one of the moving spots changed the mean estimate of its motion direction by nearly 30°, where the greatest possible change would have been 45°. This difference results from a different potency of configurational change in the two conditions. In the induced motion condition the relative displacement was between a dot and a large pattern, while in the cross experiment object-relative displacement was between two spots.

4. COMPENSATORY EYE MOVEMENT

When one turns the head and one's eyes rest on a point in the stationary environment as they usually do, eye movements take place that compensate for the angular displacement of the head. Such compensatory eye movements resemble pursuit movements in that the eyes track a point in the environment, but they are different in that they start immediately when the inception of a head movement produces a relative displacement between the environment and the eyes. Ocular pursuit, on the other hand, begins after a short delay, during which the displacement of the object that starts to move is given as an image displacement. The immediate start of the compensatory eye movements is made possible by a reflex that sets the eyes in motion as soon as the head begins to turn. This reflex can be demonstrated when a subject turns his head in the dark; his eyes will then move as if they were fixed on a stationary point.

It would, however, be a mistake to conclude that compensatory eye movements are merely reflexive and do not mediate motion perception. When one creates artificial conditions where the visual environment of a subject moves about him when he turns his head, e.g., clockwise when the head turns clockwise, the subject will perceive this environmental motion correctly, provided, of course, that it is above threshold. This threshold is larger than ordinary motion thresholds; subjects see the environment move when it revolves about them by 2 or 3 percent of the simultaneous head rotation. Yet, this threshold is amazingly small when one considers the perceptual function that makes motion perception possible when compensatory eye movements are the stimuli.

When a subject turns his head to the right, say, by 15 degrees, the environment undergoes a revolution relative to his head of 15 degrees in the direction opposite to the head movement. This rotation of the environment would be perceived were it not caused by the head rotation. Therefore, seeing the environment as stationary under these conditions, as every normal subject does, involves taking the head rotation into account. This matching up of the 15° visual displacement of the environment with the 15° head rotation must take place with great accuracy, for the threshold just mentioned means, of course, that, if the environment turned by 15.5° instead of by 15°, it would be seen to move by a majority of subjects. Similarly, motion of the environment will be perceived when one arranges to have it move less than 15° to the left when the head turns 15° to the right. Such arrangements involve making an artificial visual environment which moves dependent on head rotation. If it is made to move in the direction with the head rotation, the relative displacement between the environment and the head will be smaller than normal, and if this deviation from the normal relative displacement is above threshold, the environment will be seen to move with the head whenever it is turned.

Wallach and Kravitz (1965) first measured these thresholds and showed how accurately the process operates that matches up visual displacement of the environment with head rotation. They used an arrangement in which a mirror that turned on a vertical shaft was coupled to a helmet worn by the subject via a variable ratio transmission that enabled the experimenter to adjust the ratio between the rotation of the mirror and the head rotation. The mirror reflected a projector beam onto a screen in front of the subject, and when a rotation of the head turned the mirror, the scene before the subject shifted in any desired proportion of the head rotation, either in the direction of the head movement or in the opposite direction. Thus, this apparatus could be used to present a subject with a displacement of his visual environment that depended on his head rotation, and it enabled the experimenter to vary the proportion between the displacement of the visual environment and the head rotation by adjusting the variable ratio transmission. The apparatus yielded a reading of the proportion between the angular displacement of the environment and the head rotation which we called displacement ratio (DR).

Wallach and Kravitz used this apparatus to measure the two motion thresholds connected with the apparent immobility of the visual environment during head turning. One was the displacement ratio (DR) at which environmental displacement <u>with</u> the head rotation became noticeable, that is, where environmental displacement due to head rotation was made smaller than normal, and the other one was where the environment was made to shift in the direction against the head turning. The range of DRs between the two thresholds at which the displacement of the environment dependent on head rotation was too small to be noticed was called no-motion range. It was between .04 and .06 DRs wide and lay roughly symmetrically about the point of objective immobility.

The perceived stability of the environment during head movements, thus, is due to a process that renders ineffective the visual displacement that is caused by the head movements so that it is not perceived. Our demonstration of a small no-motion range shows that this compensation process operates only on that part of the environmental displacement that approximately matches up with proprioceptive information about the head movement, while it causes that part of the environmental displacement to be perceived as motion which is not accounted for by the head movement. When objective motion of the environment of .03 DR is added to the environmental displacement that is caused by head turning it will be perceived by most subjects. A falling short of environmental displacement by a similar amount also leads to perception of environmental motion. That is, when the environment is so moved during a head rotation that its displacement caused by the head turning is diminished, i.e., when the normal displacement is altered by -.03 DR or more, motion will also be perceived.

Evidence that the perception of such excess or shortfall displacements of the environment during head movement is mediated by compensatory eye movements emerges from a detailed investigation of the compensation process. Processes of space perception are best

investigated by subjecting them to perceptual adaptation. In the
present instance, that line of work was begun by Wallach and
Kravitz (1965) who had subjects wear minifying lenses of .66 power.
By diminishing visual angles by one-third, these lenses caused a
decrease by one-third of the visual displacement of the environment
that results from head movements. The subject who wore the lenses
saw the environment move whenever he turned his head. These optical
motions were in the direction of the head rotation and were equivalent to environmental motions of -.34 DR. After wearing these
lenses for 6 hours, all subjects saw the environment move with every
head movement. When the no-motion range was measured it was found
that its midpoint had shifted to -.175 on the DR scale (N=12). That
means that, after adapting to the lenses, subjects, on the average,
saw the environment stationary when it moved in the direction with
the head movement by 17.5% of the head rotation. After an exposure
to the lenses lasting six hours adaptation was 50% complete. (For
a more detailed explanation of this experiment see the introduction
to Wallach and Kravitz, 1968).

Eventually, we found that much shorter exposure periods produced highly reliable adaptation effects, provided the subject kept
turning his head back and forth continuously during the allotted
time. The test apparatus was used to present the environmental displacement during head turning to which the subject adapted. In most
of the experiments to be reported the visual displacement was in
the direction with the head movement and the ratio of the visual
displacement to the head rotation was .4, that is, the apparatus
was set for -.4 DR. Before and after exposure to these conditions,
the two limits of the no-motion range were quickly obtained by
gradually diminishing a displacement ratio that caused perceived
motion in one direction until the subject, who turned his head in-
cessantly back and forth, saw no motion and by repeating the procedure starting with a displacement ratio that caused motion in the
other direction. A ten minute exposure yielded an adaptive shift
of the no-motion range of about .1 DR, that is, the mean of its
midpoint was located at -.1 on the DR scale after exposure. Such
a rapid effect made an investigation of this adaptation process
possible.

5. ADAPTATION AS CHANGED EVALUATION OF COMPENSATORY EYE MOVEMENTS

What is here of special interest is the finding that an exposure in which the environmental displacement is given solely by
compensatory eye movements can lead to adaptation. This result was
obtained when we raised the question whether adaptation consists in
a changed evaluation of the eye movements that provide the information that the environment is abnormally displaced during head movements, or whether it consists in a change in the evaluation of a
representation of the environmental displacement at a higher level.
In the latter case it should not matter how the nervous system

obtains the information about the abnormal environmental displacement; adaptation should result also from an exposure to abnormal
environmental displacement during which the eyes undergo normal
compensatory movements. Such an exposure could consist of a condition where the visual environment is made to shift dependent on the
head rotation, while a stationary mark is given in the center of
the field. If the subject fixated that mark, his eyes would make
normal compensatory movements, while he would see the visual environment move with each head movement. The latter motion would
here be given by image displacement. If, on the other hand, our
adaptation depends on the subject's eyes making abnormal compensatory movements in the exposure period, a condition in which only a
fixation mark is being displaced dependent on the head rotation
while the visual environment remains stationary should lead to
adaptation. We found that both of these exposure conditions lead
to an adaptation that can be measured in the same way in which we
had always measured adaptation. We called the adaptation obtained
when the visual field is displaced while the eyes fixate a stationary mark, and therefore make normal compensatory movements, "field
adaptation". After a ten minute exposure to field adaptation conditions in which the environment was displaced at -.4 DR, Wallach and
Bacon (1977) obtained a mean change in the no-motion range of -.056
DR, which, with N=28, was highly significant (t=13.6). We called
the adaptation obtained when the visual environment remained
stationary and only the fixation mark underwent head movement-
dependent displacements "eye-movement (EM)adaptation". After a ten
minute exposure to EM adaptation conditions where the fixation mark
was displaced at -.4 DR, Wallach and Bacon measured a highly significant mean change of the no-motion range that amounted to -.072 DR
(t(27) = 11.9).

    What happens to compensatory eye movements when they partially
adapt to the head movement-dependent displacements of the fixation
mark? When, in the experiment just reported, the fixation mark moves
during exposure at a rate of .4 of the head rotation always in the
direction with the head movement, its displacement relative to the
head is 40% less than the displacement of a stationary mark. Initially, this shortfall is perceived as a motion of the fixation
mark. Presumably this perceived motion decreases as adaptation sets
in. Complete adaptation to the 40% diminished displacement of the
fixation mark would mean that the mark is perceived as stationary.
If this were solely a matter of eye movements, an overrating of the
eye movements by 40% would compensate for the 40% shortfall and
would represent the fixation mark as stationary. The partial adaptation of -.072 DR that Wallach and Bacon measured after a ten minute
exposure can be viewed as an overrating of the eye movements by
7.2%. The following experiment shows that this explanation is
correct. Before and after an eye movement adaptation exposure
identical to the one just described, Wallach and Bacon gave a test
in which the subject had to point at a target immediately after he
had made an eye movement of known extent. In total darkness the

subject turned his head to the left until a stop halted his movement at a point at which the head was turned by 18° out of the
normal position. As soon as turning of the head stopped, a vertical
line straight in front of the subject's body lit up and he had to
point to it. If his compensatory eye movements to the right were
overrated, he would be expected to point farther to the right after
the adaptation exposure than before, and this was indeed the case.
Translated into the displacement ratio scale, the mean change in
the pointing direction for 12 subjects amounted to .133 DR. This
change in the pointing direction was significantly larger than the
effect of the adaptation that had been measured as a shift of the
no-motion range after eye movement adaptation; that shift had
amounted only to .072 DR. The difference probably results because
the latter test measures not only EM adaptation but also probes for
an effect of field displacement. Since the field remains stationary
in the EM adaptation condition no field adaptation can develop. The
higher level process where field adaptation would take place
registers environmental displacements normally, while, at the same
time, the evaluation of eye movements is changed. The shift of the
no-motion range seems to depend on a combination of these two
states. The pointing test, on the other hand, only probes the state
of the eye movement system. This fact was ascertained by having
another group of 12 subjects undergo field adaptation. No change
in the pointing direction resulted.

The experiments which I have reported show that the compensation process that prevents us from seeing the environment move when
we move the head can be rapidly altered when the subject is presented with a condition where the environment moves in a regular
fashion dependent on the head rotation. Such an adaptation can consist either in a changed relation between head position and a representation of the visual environment that is so constituted that
the result of eye movements has already been taken into account
(field adaptation) or in a changed evaluation of compensatory eye
movement (EM adaptation). The finding that EM adaptation causes a
shift in the no-motion range means, of course, that a changed
evaluation of compensatory eye movements causes the stationary
environment to be perceived in motion when the head turns. Thus,
compensatory eye movements may lead to motion perception. Normally
their function is to bring about the perception of a stationary
environment; only experiments with adaptation show that they can
also cause a perception of motion.

At the beginning of the discussion of compensatory eye movements it was mentioned that they operate as a reflex to head movements. What we now know about EM adaptation shows that, like pursuit movements, compensatory eye movements mediate displacements
between a tracked point and the head. They can do that only if they
depend more on visual stimulation than on head movements. It seems
even possible that the pursuit function is primary in compensatory
eye movements, and that they become reflexes to head movements
through a learning process. Such learning would cause proprioception

representing head movements to substitute for the displacements of the visual environment. If that is the history of reflexive compensatory eye movements, it may be possible to alter them by prolonged exposure to abnormal environmental displacements during head turning. Wallach and Bacon were able to demonstrate such an adaptive change in the purely reflexive compensatory eye movements that occur in the dark.

Lacking eye movement recording equipment, they measured the eye position at the completion of a reflexive eye movement with the help of an afterimage. The subject with the head in normal position first fixated a luminous spot straight in front of him. He was instructed to turn his head to the left as soon as the spot disappeared and to keep his eyes on the place where the spot had been. A stop limited the head movement to 15°, and a switch activated a flash gun as soon as the stop was reached. The flash lit an opening in the shape of a pointer and was located just below where the fixated spot had been a fraction of a second before. A screen in front of the subject with a vertical straight line at its center was then lit. The subject was instructed to turn his head back to the normal position and to look at the line. Keeping his eyes on the line, the subject gave the experimenter instructions that enabled him to mark the spot where the subject saw the afterimage. This procedure was repeated three times and three locations of the afterimage relative to the fixated line were obtained and averaged. A ten minute period of EM adaptation to the motions of a fixation mark that was displaced at the rate of -.4 DR followed, during which the subject made compensating eye movements of diminished extent. Afterwards the subject was given a single afterimage test. After the adaptation period, the afterimage was located farther to the left than before, and this showed that the compensatory eye movement that was made in the dark was shorter after EM adaptation. Transformed into a displacement ratio, this mean shortening of the reflexive eye movements amounted to .13 DR. A week later, each of the ten subjects who participated in this experiment were given the afterimage tests before and after "field" adaptation. No change in the reflexive eye movements was found, and the difference in the results of the different adaptation conditions was significant at the .01 level.

I am inclined to infer from the rapid change of the compensatory eye movement reflex with EM adaptation that it is learned. If that is correct, compensatory eye movements are really pursuit movements that have acquired a connection with the head movements during which they regularly take place. It is this acquired connection that causes them to start without a delay when the inception of a head movement produces a relative displacement between the environment and the eyes.

REFERENCES

Brown, J.F., The visual perception of velocity. Psychologische
    Forschung, 1931, 14, 199-232.
Duncker, K., Ueber induzierte Bewegung. Psychologische Forschung,
    1929, 12, 180-259.
Johansson, G., Configurations in event perception. Almqvist and
    Wiksells Boktryckeri AB, Uppsala, 1950.
Kling, J.W., and Riggs, L.A., Experimental Psychology, Holt,
    Rinehard and Winston, Inc., 1971.
Mack, A., Fisher, C.B., and Fendrich, R., A re-examination of two-
    point induced movement. Perception & Psychophysics, 1975, 17,
    273-276.
Proffitt, D.R., Cutting, J.E., and Stier, D.M., Perception of
    wheel-generated motions. Journal of Experimental Psychology:
    Human Perception and Performance, 1979, 5, 289-302.
Wallach, H., Visual perception of motion. Chapter V In: On Percep-
    tion, H. Wallach. Quadrangle/The New York Times Book Co.,
    New York, 1976.
Wallach, H., and Bacon, J., Two kinds of adaptation in the constancy
    of visual direction and their different effects on the percep-
    tion of shape and visual direction. Perception & Psychophysics,
    1977, 21, 227-242.
Wallach, H., Bacon, J., and Schulman, P., Adaptation in motion
    perception: Alteration of induced motion. Perception & Psycho-
    physics, 1978, 24, 509-514.
Wallach, H., and Kravitz, J.H., The measurement of the constancy of
    visual direction and of its adaptation. Psychonomic Science,
    1965, 2, 217-218.
Wallach, H., and Kravitz, J., Adaptation in the constancy of visual
    direction tested by measuring the constancy of auditory direc-
    tion. Perception & Psychophysics, 1968, 4, 299-303.

VISUAL SPACE PERCEPTION THROUGH MOTION

Gunnar Johansson

University of Uppsala
Sweden

SUMMARY

Mathematical flow models for description of the optic stimulation are contrasted to the traditional static image model and a
perspective flow model of the type initially introduced by
J.J. Gibson is specifically argued for.

Till recently the experimental verification for this model has
been limited to object motion perception in central vision. In this
paper the first results from a program concerning investigation of
peripheral three-dimensional space perception are described and
their theoretical implications are discussed. In short, these studies
make clear that the retinal periphery in an astonishingly efficient
way respond to perspective flow patterns in terms of perception of
a rigid three-dimensional space.

Also the perceptual metric for veridical space perception is
discussed. Furthermore, some viewpoints on essential problem areas
for future research in visual space and motion perception are
given.

INTRODUCTION

A couple of years ago I saw a wonderful German film showing
the visual capacity of the male house fly during the mating procedure. The flying couple performed a sort of irregular dance where
the female part circled around in a randomlike path while the male
fly followed the female in pace describing the same course some
inches above the female. What an excellent visual achievement of
the male part! Surely no military pilot could "lock" to another
plane in a more perfect way than this fly. Evidently, a complicated

19

neural structure of cortical type like that in the higher vertebrates is not a prerequisite for excellent visual performance.

Similar types of mating behavior is well known as typical also
for birds and we also know from experimental research that newly
hatched chickens can visually orient themselves and with good precision pick grains from the ground. Furthermore, when the experimenter by means of prisms displace the foveal optic flow relative
to the retina the animal in accordance with the displaced image
picks beside the grain.

From these and many similar examples we must draw the conclusion that early in the evolution very simple and spontaneously
functioning but still highly qualified and effective visual systems
were developed and that, at least in some vertebrate organisms with
eyes of the same type as in man, visual experience is not needed
for efficient visuomotor performance. Evidently the traditionally
assumed cognitive processing of the two-dimensional optic input to
the eye is not a sine qua non for efficient vision neither in
arthropods nor in vertebrates. And if - as earlier sometimes was
assumed - the visual system in man were incapable of handling
visual information without some "higher" type of problem solving
activity this must imply a strange degeneration of the visual
system somewhere in the history of evolution between bird and man.
However, recent research on visuomotor control in infants also
indicates that there exists in man as well as in birds an ability
to perceive the three-dimensional space in a spontaneous and direct
way. (For recent material and references about infant vision, see
Johansson et al., 1980).

The classical indirect-perception theory of Helmholtz' type
with experience, learning and unconscious cognitive processing of
sensations as fundaments for visual performance was founded on the
philosophical doctrine of empiricism and in experiments where the
stimuli mainly were static artefacts. This theory put the burden
of adaptation of the visual system to the environment on the individual organism rather than on the evolutionary development of
the species. The insight of today about the neural preprogramming
of the visual systems, for instance when it is a question of recording environmental motion, has given support to a more direct
type of theory and so have the studies on motion perception in
man.

Personally, I regard the traditional polarization between
direct and indirect theories in this field as rather irrelevant.
My background interest in biology and my experimental research on
motion perception have made me convinced that also in man like in
the fly and bird we must assume a neural "hardware" for visual
stimulus processing, the structuring of which is genetically
encoded. Certainly this hardware is not less efficient in man than
it is in the fly, the frog or the cat but because of the more complicated CNS in man our visual adaptability is far higher. Therefore we ought to make a distinction between basic space perception
and, for instance, perception of human-made artefacts like drawings,

paintings, etc. The very basic function for vision biologically spoken, is to help the organism to adapt its behavior to the physical environment. In an essential way it concerns detecting food and dangers together with control of locomotion for escape or for searching or hunting food. Thus, action (active selfmotion) and visual perception are interrelated in a most basic way.

At least in man (an in some primates) there surely exist also a capacity of visual problem solving as an elementary type of cognitive analysis. For instance the capacity of "reading" simplified static pictures seems to a certain extent to be a cultural product. Therefore, the either-or type of discussion hardly can be meaningful. More interesting is to ask for primary and secondary functions and to ask what functions are possible to modify. Here again the motion perception and action aspect affords an interesting example: we cannot perceive motion in three-dimensional space as a two-dimensional spatial change in accordance with the retinal flow but we readily can recalibrate optically induced distortions of the flow pattern. Every occasional user of positive eyeglasses makes such a recalibration each time he puts on or takes off the glasses. Optically the floor or table top is closer with the glasses than without but the active movements in the visual space always are perfectly well adapted to the "real" distance. The studies of the slow adaptation to distortion of the visual input with prisms and mirrors of Erisman-Kohler type, afford other fascinating examples of far-reaching visual ability to modification and Wallach has demonstrated the existence of far more rapid modifications of the same type which not look so dramatic but which from theoretical point of view are most informative (Wallach, 1976).

However, my interest in the present paper mainly concerns the basic non-adaptable component. Let us therefore return to the basic relation between action and motion perception. From point of view of perception this gives us two essential categories of motion perception: perception of objects (e.g. food), stationary or moving, and perception of the organism's own active motion relative to the ground, thus locomotion or self-motion. In the following I will treat both of these aspects of visual control of the organism's behavior in the three-dimensional space. From point of view of stimulus description there exists a difference between these two types of motion perception. Self-motion perception typically is an effect of a mathematically coordinated stimulation of most parts of the retina. In the following this will be termed "wide angle stimulation. Object motion usually covers rather limited visual angles. For this latter type of coordinated stimulation I will in the following use the term "narrow angle stimulation". We will also observe perception of passive self motion which can be regarded as a special category between the two mentioned.

The question about accepting or rejecting the possibility of sensory processing of the proximal stimulus has a special relevance in our present context. If we should be enforced to assume that the transfer from optic stimulation to space perception were of a

problem solving type, we hardly could expect any direct mathematical
relationship of one-to-one type between stimulus and percept. On
the other side, if we can presuppose a direct and automatic neural
processing in the visual sensorium we must expect that establishing
such a mathematical relationship will be possible and furthermore
expect that the mathematical stimulus correlate to the percept must
be a correlate also to the neurophysiological processing. This
latter statement is of deciding importance for the possibility of
unifying the neurophysiology and perceptual psychology; thus to use
knowledge gained in one of these branches of visual reseach to
make predictions in the other one.

1. PROXIMAL STIMULUS AS AN OPTIC FLOW

For our pioneers the analogy between the camera obscura and
the vertebrate eye was of great importance. Starting from this
analogy they described the optic stimulation of the retina as a
static two-dimensional image like the image on the back wall of
the camera. With this model as a reference frame they formulated
some basic problems for visual perception and therefore these
also essentially concerned static images of camera obscura type.
When motion perception was dealt with in this framework it consequently was treated mainly as an effect of sequences of static
images.

In the advancement of perceptual research this framework and
this approach have had a serious blocking effect and have raised a
number of unnecessary problems especially in connection with object
constancy (Johansson, 1977a). A main deficiency in this approach is
that the analysis is static, which means that the two spatial
dimensions of the optic projection onto the retina are taken into
account but not the dimension of time. In real life the nodal point
of the eye typically is in motion relative to the geographic environment, thus causing a continuous change in the retinal projection
(even in rest the head usually is swaying more or less). Furthermore, this environment itself contains objects in more or less
complex motion, it may be other organisms, vegetation, vehicles,
etc. Therefore a more sophisticated description of the optic im-
pingement must include also change along the time axis together
with the two spatial dimensions of the projections. We must study
the transformation pattern rather than a snapshot (Gibson, 1950,
1957, 1966; Johansson, 1975, 1977a, 1978a). In this way we get a
model with a two-dimensional flow representing rigid motions in
three dimensions. The increasing interest in motion perception
research (or more generally: event perception research), of which
the present symposium is a good indication, has actualized also
the flow type of stimulus description.

The perception of motion traditionally was treated as a
troublesome special case of perception. My arguments above about
perception of self motion and object motion as most common compo-

nents in space perception states the opposite: motion perception
must be regarded as the general case. I will in the following make
clear that also what has been called static perception in fact
essentially is a product of motion perception, namely self motion
perception.

By definition a flow represents some kind of non-rigid motion.
Streaming motion in fluids or gases are typical examples. As is well
known it most often is a hard problem to develop mathematical models
for such events. In this respect, however, the non-rigid optic flow
over the retina is advantageous. It has only two spatial dimensions
and it is the effect of a central projection through the nodal point
of the visual lens system. Furthermore, to the extent it is generated by motion of rigid structures, it is a perspective projection and
thus mathematically lawful. Therefore, if rigidity is existing distally, this optic flow can be specified mathematically in various
ways. This has also been done in several analyses (Gibson et al.,
1955; Hay, 1966; Lee, 1974; Gordon, 1965; Nakayama & Loomis, 1974;
Prazdny, 1980; Ullman, 1978). These types of analyses are in a
direct (Gibson, Hay, Lee) or indirect way related to a description
in terms of perspective transformation.

Gibson (1957) explicitly proposed perspective transformations
as a model for perception of rigid motion perception. In my own
research I have also found that a model based on some fundamental
relations in projective geometry can form an adequate and highly
convenient framework for description of stimulus flow-motion perception relations and I will deliberately apply it also here.

1.1. About experimental policy and terminology

One of the most basic problems for research on visual space
and motion perception can be expressed in the following form: what
factor in the proximal stimulus flow determines the percept?
Therefore, in the experimental research it is advantageous to
choose a mathematically well defined (or definable) optic flow as
the stimulus. The experimental task then is to get a description
of the percept which can be handled in some kind of geometric
terms. Finally, the experimenter must search (most often in a
stepwise way) for a mathematical transformation of the stimulus
description which as closely as possible corresponds to the percept description. I have programmatically worked in this vein and
therefore I am very particular about choice of stimulus as well as
specification of the percept in analyzable terms. Therefore, before
proceeding, I will define some essential terms to be used in my
description of the percepts. Perceptual terms to be used:
Environment: visual representation of the space as contrasting
to the own body.
Self motion: perceived motion of the own body, head and/or
eyes relative to the environment.
Stationary or static objects: objects perceived as stationary

relative to the environment.

   Object motion: perceived motion of objects relative to the
environment.

2. ABOUT THE PERCEPTUAL VECTOR ANALYSIS AS DEMONSTRATED IN OBJECT MOTION RESEARCH

   Thirty years ago I demonstrated in a series of experiments the
basic principle in the perceptual treatment of the proximal stimulus
flow which is called the perceptual vector analysis (Johansson,
1950). I will presuppose that this effect in its most simple form
as illustrated in Figure 1a is known for most of my readers.

   Already this simple example says that generally the optic flow
is not perceived as it appears on the retina or as experimentally
presented on a frontoparallel screen. Instead it is perceptually
analysed in a number of components abstracted from the total flow
pattern. Each of these components represents a motion relative to
the component generated by the self motion (thus the component representing the perceptual environment), or relative to some of the
other motion components. It is not hard to realize that in fact some
kind of neural processing analogous to such a mathematical vector
analysis of the optic flow is necessary for perceiving an environment with self motion and object motion. Each perceived motion must
be assumed to have some kind of neural processing counterpart.

   The example in Figure 1a is simple not only because it has
only two elements in motion against a stationary background; it is
simple also because it represents a situation with frontoparallel
motion tracks. In this respect Figure 1c is more interesting because
it is an example of a perceptual motion analysis of elements moving
in depth with Figure 1b representing the frontoparallel projection
of the same pattern. In this example (where the combination in 1b
is well known) the element H, when seen alone, usually is perceived
as moving in the frontoparallel plane while the distorted cydoidal
motion of R is seen as a kind of jumping or bouncing in depth. Exposed together the two elements are perceived as a rolling in depth.

   This example introduces the use of the projective model.
Treated as perspective projection these two examples (Figure 1b and
c) are equivalent in the meaning that by an appropriate perspective
transformation Figure 1b can be transformed to Figure 1c and vice
versa. With other words: each non-rigid perspective projection of
a rigid motion can be transformed to its rigid equivalent. I have
previously used this transformation to rigid motion of a component
in the optic flow as a model for the perceptual processing of a
proximal non-rigid perspective transformation and used the term
"reverse projection" for this process (see Johansson, 1964, 1980).
Because of its special geometrical simplicity, this special case in
the family of equivalent perspective projections of a figure can
always be specified geometrically. This implies that, for instance,
combinations of motions in a plane like those examplified above can

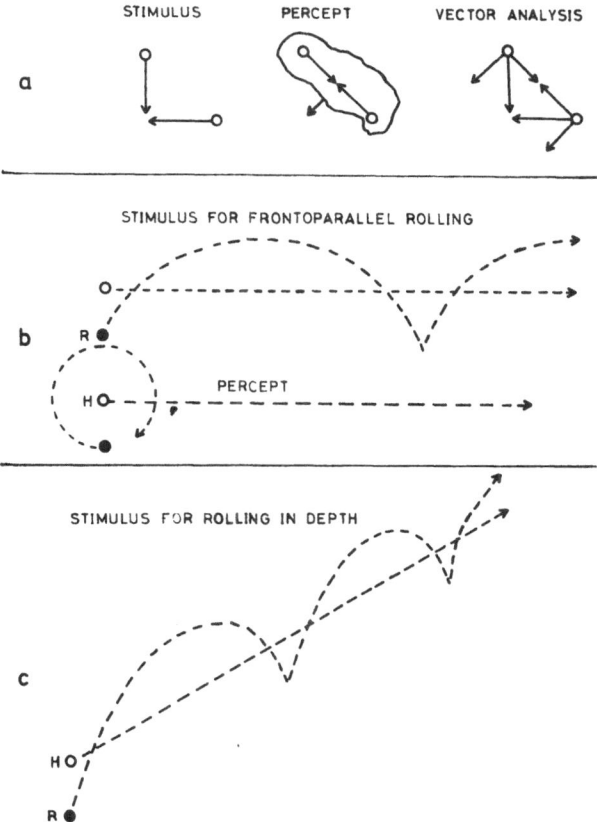

STIMULUS   PERCEPT   VECTOR ANALYSIS

a

STIMULUS FOR FRONTOPARALLEL ROLLING

b

PERCEPT

STIMULUS FOR ROLLING IN DEPTH

c

Fig. 1. Three examples on perceptual vector analysis: (a) Example
on the basic analysis in common and relative motion components. (b) The frontoparallel case of perceptual abstraction of a translatory component in cycloidal motion common
with a translatory motion. (c) The same patterns as in (b)
but projected from another angle.

be transformed to this simple form and analysed in terms of common
and relative rigid motions.

The examples discussed so far are simple also because they
exemplify the case with only one common motion component, the
simplest form of hierarchy between common and relative motions.
Figure 2 is an example of a far higher degree of complexity in this
respect: the perceptual resolving of the also distally non-rigid
motion of the human body (represented by one element on each of the
main joints) as a hierarchical system of rigid pendulum motions.
Here we get a complicated hierarchy of common and relative motions
which in a most vivid way is perceived as a walking human being as

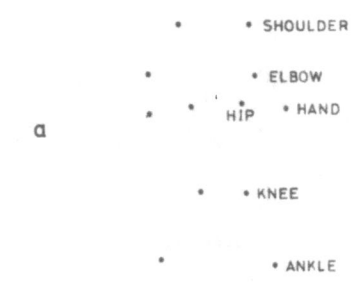

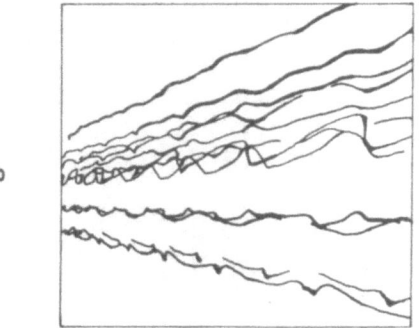

Fig. 2. Example on biological motion analysis. (a) The arrangement
        with light spots on the main joints of a walking man. The
        motions of the spots were recorded (by film or video) against
        a totally dark background. (b) An "open shutter" record
        of the motion flow.

shown in Figure 2. The approximately translatory motion of the hip
during walking forms a common component for all joint motions, the
knee element is seen as moving relative to this common motion but
it also forms the reference frame for the perceived ankle motion,
etc. Many experiments with various types of such 3-D "biological
motion" patterns sometimes with common mechanical components added
have made clear that also highly complex hierarchical systems of
common and relative motions are immediately analysed in the way
described. For a vector analysis of such a pattern in correspondence
with the perceptual outcome see Johansson (1976).

      I have used the terms simple and complex above in the tradi-
tional way as related to the geometrical description. From point of
view of perceptual analysis this may be rather misleading. The
motion combinations are regarded as simple or complex from our
mathematical or descriptive point of view. However, I have found
that the following rule has a very good general validity: stimulus
flow patterns which are simple to describe mathematically, generally

are hard and artificial for the visual system to handle in a specific way while patterns which seem most simple and self-evident
in the visual processing most often are extremely complicated and
hard to specify mathematically. Evidently the mathematics of the
neural processing of the optical information has another character
than our mathematical methods. Probably a kind of parallel processing makes the system most efficient only when the amount of
information is very rich and redundant.

So far the perceptual vector analysis has been studied only in
cases of object motion. In these studies it has been found to be
mainly a narrow angle effect. The full coherence effect between
moving elements seems to be limited to a few degrees of visual
angle. Gogel has in an experimentally elegant way demonstrated this
characteristic (Gogel, 1974). It is tempting to hypothesize that
this in some way neurologically should be related to the size and
organization of the receptive fields. We may remember, however, that
this limitation is observable only under experimental conditions
with impoverished information like those described above. In everyday perception the system usually is dealing with optic projections
of structured objects with a tremendous amount of information-
carrying optic elements. And in accordance with the rule just given,
this gives a good continuity and a highly specific perception.

2.1. Perception of self-motion and a stationary environment

Perception of self-motion and of the visual environment are
closely related. They usually are effects of the same retinal stimulation, namely the coherent background flow generated by motion of
the perceiver's head relative to the physical environment. It goes
without saying that these aspects of visual perception are of
essential interest for the structuring of a comprehensive theory
of visual space and motion perception and I will consider both
these aspects, beginning with the self-motion effect.

Self-motion perception, after being more or less neglected
since Mach's initial studies, has attracted a considerable interest
during the last decade, especially among students of sensory
physiology. A good indication of this renewed interest is that a
whole section in the present symposium has been devoted to this
aspect of motion perception. Therefore I have good reasons to limit
myself on this point to summarizing some basic findings of special
interest in the present context. For further references I will
just mention Dichgans and Brandt (1978) and Johansson et al.
(1980).

One such finding which seems to be something like a common
denominator in the research on self-motion is that a moving stimulus
pattern projected onto the peripheral parts of the retina of a subject in rest, elicits in most subjects a highly realistic impression of self-motion (passive self-motion) while the same stimulation
over the central parts (here defined as an area with a radius less

than 15° from the fovea) never gives this effect but instead is
perceived as an object motion. This says that (1) the central part
of the retina and the peripheral parts responds in mutually different ways to the same stimulation and (2) that the periphery is
so tuned for self-motion interpretation of moving stimulus patterns
that this perception is evoked also in many stationary subjects,
yielding passive self-motion perception in lack of e.g. vestibular
motion reactions. We also can find that it is the peripheral self-
motion interpretation which determines the percept when both central
and peripheral parts are uniformly stimulated by motion.

Under the artificial conditions described with a stationary
perceiver and moving stimulus pattern not all subjects perceive
self-motion and a latency period is typical. However, Lee and his
co-workers in their studies of the "swinging room" effects have
found that the effect can be totally general and immediate under
their more "realistic" experimental conditions (Lee & Aronsson,
1974).

The experiments referred to so far are experiments with wide
angle stimulation of the periphery. Is the self-motion response an
effect of stimulation of the retinal periphery or just an effect
of wide angle stimulation? The answer to this question is that it
is the retinal locus and not the extension of the stimulus area
which is the deciding factor. I have been able to establish and
quantify passive monocular self-motion perception from a single
bright dot in slow oscillation 75° peripherally from the fovea and
with peak to peak amplitudes from about 3° to 22° visual angle
(Johansson, 1977b; cf. also Johansson, 1977c).

2.2. Wide angle stimulation and perception of the environment

Evidently the peripheral part of the retina in an effective
way reacts with self-motion perception in spite of the low degree
of resolution. Can it also bring about valid perception of the
three-dimensional space? Let us set this problem in the following
form. As pointed to above, a motion (passive or active) of the
perceiver's head relative to the rigid environment generates an
optic flow over the whole retina which generally is non-rigid but
which also always represents a coherent set of perspective transformations. By an appropriate reverse projection such a flow always
can be transformed to a rigid motion in Euclidean meaning, thus a
rigid structure in relative motion. My question is: can the visual
system really make use of the information inherent in a perspective
flow over the retinal periphery? Can it restructure the flow in a
way analogous to a projection to rigid form?

This type of questions has occupied me a lot during the last
years. Practically all experimental studies on the effects of perspective transformations in the optic flow have been limited to
central vision and therefore I have begun some frontal attacks to the
wide angle problem. In the tradition of static perception studies

my problem was of limited interest. We have been learned that the peripheral stimulation with all its blurredness (especially under static conditions) is of little use for space perception but because of an automatic reflex it is relatively efficient for detecting sudden motions. Therfore, I must admit that in spite of my own anchorage in event perception I was hesistant about the capacity of the peripheral retina in the respect discussed. The geometric achievement necessary in combination with the low degree of resolution available, a priori seems to make the situation very problematic.

The experiments carried out so far, however, have in a most convincing way refuted my initial scepticism. The studies so far have shown that peripheral wide angle perception can bring about a highly specific and relevant perception of a three-dimensional environment. I will in the following briefly describe the experimental conditions in these experiments and some more or less preliminary results and observations obtained so far. (The first detailed report on these studies will be published in the near future. Some of the pilot studies are described in a prepublication (Johansson, 1980).

Two types of display have been studied so far. Both represent a proximal flow on a vertical screen and have the form of a perspective transformation as generated by projection from a rigid motion over a horizontal plane surface. One of these patterns consists of the projection from a set of parallel straight lines and the other of the projection from a square with 100 randomly distributed dots, both in linear motion. Figure 3 illustrates these two types of display.

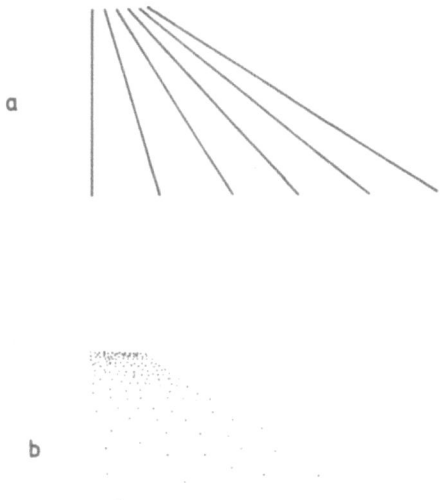

Fig. 3. Two types of displays representing wide angle projections of linear motion over a square. See text.

The display events were computed as extreme wide angle projections (79° x 72°) and presented to the subject on a vertical sagit-
tal-parallel screen, monocularly viewed and with the active eye
exactly in the computed station point of the projection. Figure 4
shows the geometrical conditions for the computation and displaying.

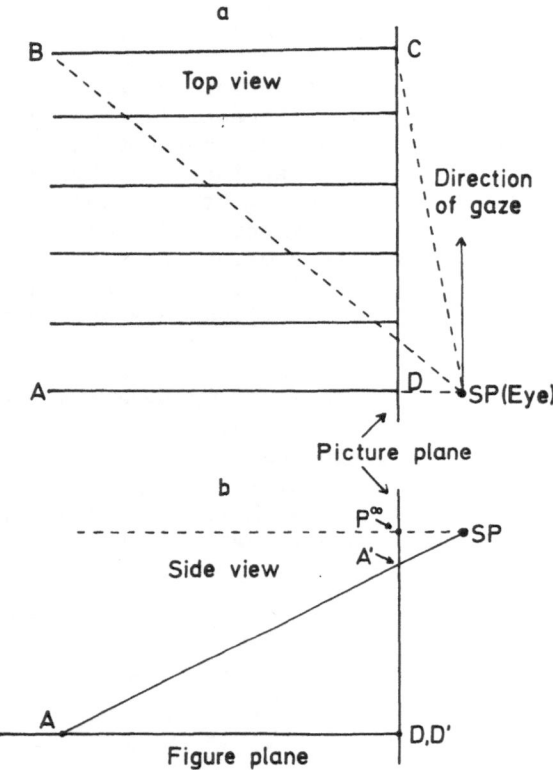

Fig. 4. Top and side views of the projection. Also the experimental
        arrangement with the eye at the computed station point and
        the gaze directed parallel with the projection screen
        (picture plane) is shown. Under these conditions the pro-
        jection onto the picture plane of the square ABCD corre-
        sponds to Figure 3a.

Under the conditions described, the continuously changing
patterns on the screen are by all observers immediately seen as a
rigid motion over a horizontal, approximately squareshaped surface,
similar to a moving endless belt. Thus, the lines are seen as
parallel and the spots as randomly distributed. This perception of
horizontal orientation, rigidity and right-angleness has a very

"real" character and allows no alternative interpretations.

It is also very interesting from a theoretical point of view to find that, when the line pattern is kept stationary under the experimental conditions, it still gives a good impression of horizontal orientation and parallelity. This impression is lacking in the dot pattern. Thus, the line pattern studied makes possible a static perspective analysis, but the density gradient in the point pattern is not effective in this situation.

The experiments described, indicate that the retinal periphery spontaneously processes the wide-angle proximal flow in accordance with the principles of invariance under perspective transformations. The low degree of resolution power in the optic-retinal system seems not seriously to interfer with the possibility to pick up relevant information. However, some still more impressive indications of the visual sensitivity to the perspective projection and rigid motion information available in a perspective transformation, were found in this set of experiments. Contrary to the outcome from the narrow-angle type of experiments it was namely found that the *exact polarity of the projection is decisive for the perception under our wide-angle conditions*. As I have underlined elsewhere (1978b) under narrow-angle conditions the visual system is in a strange way insensitive to the distance and angles between the display with a perspective transformation or picture and the perceiver's eye. This makes it possible for instance to watch TV from most viewing points in a room and in the same way to adequately interprete perspective pictures and photographs. In the wide-angle experiments described it was instead found that the least displacement of the eye relative to the computed station point brings about specific and geometrically predictable changes in the percept. Also small displacements of the eye in vertical direction results in a geometrically lawful change of slant of the perceived surface. In an analogous way a small change from the correct distance eye-to-screen in the same predictable way changes the perceived extension in depth of the perceived surface. Figure 5 illustrates the geometrical lawfulness underlying these effects.

At least from my own theoretical point of view such studies of wide-angle perception can to a considerable extent widen our knowledge about the visual processing of the stimulus flow. Our traditional extrapolation from what is known about central vision in this respect can be seriously misleading. The results reported here certainly indicate – albeit so far in a somewhat preliminary way – that the visual system as a unit is highly sensitive to the lawfulness in perspective transformation and that this effect is unique for wide-angle presentation.

2.3. About absolute visual measures of the environment

As has been observed the spatial information inherent in the optic flow over the retina is of a relational character. It contains no information about the geographical distances and direction, only

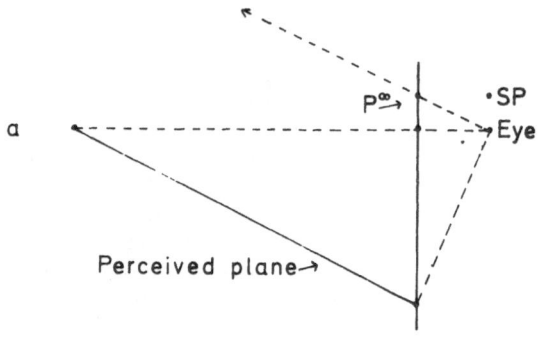

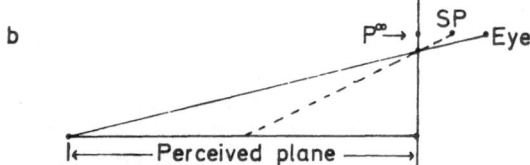

Figure 5. Reverse projection from the actual locus of the eye to a
plane where the projected motion is a rigid motion. (a)
shows a case with the eye vertically below the station
point of the proximal pattern and (b) with the eye horizontally displaced from the station point. (a) shows how
in a reverse projection from the eye a line from the eye
through the point at infinity of the projection determines the slant of the perceived surface. This is a consequence of the fact that, in a perspective projection
of a plane, the line from a point at infinity of this
plane through the station point always is parallel to the
plane. Figure (b) illustrates how the distance eye-to-
picture plane determines the extension in depth of the
reverse-projected surface.

about their relations. However, when animals and men move in their
visually recorded environment they certainly - as underlined
above - in a perfectly effective way adapt their movements to
the geographical environmemt. Therefore we must conclude that
spatial vision after all works efficiently in a metric way such
that it can guide the organism's movements in its geographical
environment, for instance by estimating geographical distances and
velocities. It is evident that, from the theoretical standpoint
advocated for in the present paper, we must hypothesize a completion

of the optical information with some kind of quantifying references in order to include these facts in the theoretical framework. The question about the sources and functioning of such a completion is of great importance for a full understanding of visual space perception. Therefore, before ending my treatment on environment perception I will briefly sketch the type of solution which I regard as most probable, the components of which by no means are original.

Basically I will find the solution also to this problem in the interrelationship between environment perception and action which forms a major theme in my present theoretical approach; the relations between the perceiver's action as a physical object in the physical world and the perspective optic flow over his/her retina evoked by this action. In short, nonvisual (or better non-retinal) quantitative recording of physical forces and the organism's own movements are interacting with the visual system and furthermore the dimension of the own skeleton can be thought of as an important unit for perceptual measures of distance. I will observe just one example crucial for understanding of the perception of absolute slant as illustrated in Figure 5a.

The geographical vertical-horizontal coordinate system established by the direction of the gravitational force has a direct representation in vision as an absolute framework for visual space perception. The direction of gravitation normally corresponds to the perceptual vertical direction. Adding a centrifugal component at an angle to this force results in a deviation of the visual vertical from the direction of gravity. There exists a number of relevant studies (see Clark and Graybiel, 1968) about perception of the vertical under experimental changes of the direction of the gravitational force. These experiments have revealed that somatosensory as well as vestibular (otolith) information about the direction of the force contributes to perception of the visual horizontal and that the subject's own active adjustments are of importance. For our present problem it is sufficient to state that the absolute coordinate axes in visual space perception is an effect of sensory interaction where under normal conditions vestibular signals probably play an important role. Spatial vision certainly acts as if the otolith organs were built-in into the visual system itself and thus yielding an absolute vertical-horizontal reference frame.

3. ABOUT PERCEPTION OF THE ENVIRONMENT AS A FUNDAMENTAL FRAME OF
   REFERENCE FOR VISUAL PERCEPTION

In the present paper I have paid much attention to the influence of peripheral wide angle stimulation on the perception of the stationary environment. In my opinion this area is well worth much attention in the future; I hope my reasons for this opinion ought to be understandable by now. One general reason of course is that this aspect of visual space perception always has been very scantily studied experimentally in spite of the fact that it plays

an important role in our daily life. Another more personal reason
for me to engage myself in such a study is that, in my own theoretical approach, the questions about perceptual mechanisms behind environment perception have more and more tended to be of key character. More experimentally gained insights in this field hopefully
will open new and more relevant ways toward a better understanding
of the general problem of space perception. Let me develop this
point a little further by using some of the findings discussed above.

There has been brought forward reasons for the statement that
the wide angle optic flow generated at the nodal point of the eyes
by motion of the perceiver perceptually evokes not a perception of
external motions but of self motion relative to a stationary and
rigid structure. All this relative motion between the nodal point
and the optic elements projected onto the retina is assigned to the
nodal point and the optic elements act as a frame of reference. So
far we hardly need the vector analysis model sketched above also if
it is a correct framework for the separation of the two components
in relative motion. However, as we have found, just this processing
of the flow generated by self-motion not only established the reference base for self-motion perception but also for all object
motions. Furthermore, it has an anchorage in the direction of the
gravitational force. Optic flow components, generated by distal
object motions, which are superimposed upon the self-motion flow
also bring about perception of motion relative to this basic flow.
Therefore, the perceiving organism (more exact: the nodal points of
its eyes) and the object are both seen as moving relative to the same
stationary framework: the environment. This certainly is a very useful solution both for man and animals! It is of vital importance both
for predator and prey in the animal world and not less important for
our own physical activity.

This locomotion-generated general reference also nicely makes
consequent the experimental finding in perceptual vector analysis
research that the "stationary" environment always forms the lowest
level in the hierarchies of relative motions commented on above.

However, it has still another interesting implication which
may be some readers initially will find rather strange. It says
that a stimulus structure which we always in a self-evident way have
regarded as static and described in terms of unchanging images instead must be regarded as an effect of peripheral abstracting of
invariants in a continuous flow generated by the perceivers own
motions. Thus, static perception in fact is generated by abstracting a common component in a flow analysis.

4. CONCLUDING REMARKS

A complete theory of visual space perception still is in the
future. In my opinion the static-image approach led us into blind
alleys and delayed the progress in essential respects and therefore
there still remains a lot of basic paradigmatic research. It is

especially the research in motion perception which initially actualized the need for more adequate models and of course my own
viewpoints on perceptional theory stems from this area. Since long
I have been engaged in motion perception research and already my
initial study of event perception enforced me to leave the traditional lines. Especially the early brilliant work of Duncker and
Rubin gave important starting impulses in a new direction and later
on some trends in the early Gibson's program have helped me very
much.

In the present paper I have sketched my standpoint of today
when it is a question of some main components in an efficient paradigm for visual space and motion perception. I have here in a programmatic way widened the problem of visual motion perception by
demonstrating that it in fact intimately concerns space perception
in general and that it can open new ways to a better understanding
of also static space perception. The reason is that what is termed
perceptual static space from the point of view of an underlying
stimulus, is represented by an effect of perspective invariance in
the optic flow. To a certain extent this turns the traditional
approach upside down. Static perception is described as an aspect
of motion perception, while traditionally motion perception was.
analyzed in terms of series of static images.

The programmatic acceptance of the retinal flow as valid type
of description for all types of stimuli for space perception also
implies a rejection of the traditional axiom about proximal stimuli
as two-dimensional. The classical statement that one of the three
spatial dimensions is lost in the optic projection of course holds
true. However, it is an equally relevant fact that we have got the
dimension of time as a new third dimension when describing the
visual decoding of the proximal stimulus as a flow analysis. In this
model the stimulus for perception of the three-dimensional world
has not the underdetermined character it was thought to have in the
traditional type of static analysis and thus it can be analyzed
mathematically instead of in terms of a number of qualitative cues.

I have also found it necessary to violate another traditional
theoretical constraint in perceptual theory: the receptor type of
definition of the senses. This, however, is not hard to give good
arguments for today and especially not in the field of vision where
traditionally "non-visual cues" have been accepted. Exaggerating a
little I have said that the visual system functions as if the
otolith organe were built-in in this system.

Acccepting the principle of sensory interaction is also the
background for another main theme in this paper: the essential role
played by intentional muscular activity in the visual establishing
of the environment as a framework for self-motion.

When the proximal stimulus is analyzed in terms of an optic
flow instead of in terms of images, some kind of structuring like
the principle of perceptual vector analysis must be regarded as a
sine qua non. Only with stimulus descriptions of this type we can
cope with the complexity of motion perception. However, I also will

underline that in its present state this principle mainly is a
formulation of some experimentally abstracted principles. In this
field much work of great theoretical importance is needed.

As I already have underlined, a general theory of space perception still is in the future. However, today we can observe some
promising trends in the actual research which certainly will contribute to important new background knowledge. I will finally point
to three such sectors. First, it is evident that the interest in
event perception (which is a broader concept than motion perception)
is rapidly growing, especially among the young perceptionists. A
number of highly interesting contributions in this field have been
published during the last few years. For references and a review
of the development see Johansson et al. (1980).

Another area of growing interest concerns the selective character of visual space perception. In daily life visual perception
essentially means scanning activity and fixation. In contrast to the
stimulus decoding process these activities certainly are of a non-
automatic character and under motivational-cognitive control. Fifty
years ago "attention" was a not accepted term for perceptionists.
It has been too much in use for "explaining" what was not understood
in the perceptual process. Today the situation is very different.
"Attention", "vigilance" etc. are very actual subjects for research.
Also in the theory of general visual perception this selectivity
function must be accepted as being of fundamental importance. We
have still today, probably due to the inheritance metioned, a very
artificial mechanistic formulation of many problems. Thus: more
research on perceptual selectivity in the future!

Intentionality and selectivity are closely related concepts.
Therefore Gibson's recent stressing of "affordance" as an important
characteristic of distal objects etc., have a direct bearing on the
present discussion, see Gibson (1979). Intentional selection is a
term characterizing the functioning of the organism while "affordance" denotes the same phenomenon when attributed to the distal
world.

The third theme for future research in space perception for
which I am very hopeful, concerns the spatial information available
in the distribution of brightness and color. Classically space perception mainly was investigated in terms of the distribution of
contours, often in line drawings. I myself have to a great extent
studied effects of changing proximal forms as indicated by contour
lines or effects of spatial relations between moving points yielding perception of "invisible" contours.

This, however, is just one aspect of spatial 3-D information
available in the optic flow.

A second aspect is the proximal brightness distribution in
form of effects of illumination of 3-D structures and most probably
this is a very important source of information. Beside the traditional type of research in terms of color constancy there has recently appeared a new type of problem settings in this field. We
may, as a background, remember the Cornsweet-O'Brien effect

(Cornsweet, 1979), Land's studies of illumination perception (Land, 1971) and the puzzling mould-cast effects from directed light (von Fieandt, 1949; Gregory, 1966).

What I in a more direct way will point to as highly promising new approaches is Bergström's (1977, 1980) original type of stimulus analysis and experimental approach and Gilchrists (1979) closely related and ingenious investigations.

This new type of research in fact implies a kind of component analysis of the proximal stimulus which theoretically is akin to the perceptual motion vector analysis. It will make it possible in the future to treat perceptual space and motion perception as the outcome of the combined information in the distribution of luminance and the contour pattern in proximal stimuli. In the same way as I here have described the perception of the stationary environment (as the effect of a common motion component in the optic flow) Bergström seems to be able to show that a common color factor representing illumination is automatically abstracted in the proximal distribution of light. He makes probable that the visual system analyses the luminance gradient in proximal stimuli in a series of components: a common component informing about illumination, a 3-D form component and a constant-color component.

REFERENCES

Bergström, S.S., Common and relative components of reflected light as information about illumination, colour and three-dimensional form of objects. Scand. J. Psychol., 1977, 18, 180-186.
Bergström, S.S., Illumination, colour and 3-D form. The Abano seminar on perception, 1980 (manuscript).
Clark, B. and Graybiel, A., Influence of contact on the perception of the oculogravic illusion. Acta Oto-Laryng., (Stockholm), 1968, 65, 373-380.
Cornsweet, T.N., Visual Perception. Academic Press, New York, 1970.
Dichgans, J. and Brandt, T., Visual-vestibular interaction: Effects on self-motion perception and postural control. Handbook of Sensory Physiology, (Eds. R. Held, H.W. Leibowitz and M.H. Teuber), Berlin, Heidelberg, New York: Springer, 1978.
Fieandt, K., von, Das phänomenologische Problem von Licht und Schatten. Acta Psychologica, 1949, 6, 337-357.
Gibson, J.J., The Perception of the visual world. Boston: Houghton Miffling, 1950.
Gibson, J.J., Optical motions and transformations as stimuli for visual perception. Psychological Review, 1957, 64, 288-295.
Gibson, J.J., The senses considered as perceptual systems. New York, Boston: Houghton Mifflin, 1966.
Gibson, J.J., The ecological approach to visual perception. New York, Boston: Houghton Mifflin, 1979.
Gibson, J.J., Olumn, P. and Rosenblatt, F., Parallax and perspective during aircraft landings. Am. J. Psychol., 1955, 68, 372-385.

Gilchrist, A.L., The perception of surface blacks and whites. Scientific American, 1979, 240, 112-124.

Gogel, W.C., Relative motion and the adjacency principle. Quarterly Journal of Experimental Psychology, 1974, 14, 425-437.

Gordon, D.A., Static and dynamic visual fields in human space perception. J. of Optical Soc. of Amer., 1965, 55, 1296-1303.

Gregory, R.L., Eye and Brain. McGraw-Hill, New York, Toronto, 1966.

Hay, J.C., Optical motions and space perception: An extension of Gibson's analysis. Psychol. Review, 1966, 73, 550-565.

Johansson, G., Configurations in Event Perception. Uppsala: Almqvist and Wiksell, 1950.

Johansson, G., Perception of motion and changing form. Scandinavian Journal of Psychology, 1964, 5, 181-208.

Johansson, G., Visual motion perception. Scientific American, June 1975, 76-78.

Johansson, G., Spatio-temporal differentiation and integration in visual motion perception. Psychological Research, 1976, 38, 379-393.

Johansson, G., Spatial constancy and motion in visual perception. In: Epstein (ed.) Stability and Constancy in Visual Perception. Wiley, New York, 1977a.

Johansson, G., Visual perception of locomotion elicited and controlled by a bright spot moving in the periphery of the visual field. Department of Psychology, Univ. of Uppsala, Sweden, Uppsala Psychological Reports No. 210, 1977b.

Johansson, G., Studies on visual perception of locomotion. Perception, 1977c, 6, 365-376.

Johansson, G., About the geometry underlying spontaneous visual decoding of the optical message. In: F.L.J. Leeuwenberg and H.F.J. Buffart (eds.), Formal theories of visual perception. Wiley, New York, 1978a.

Johansson, G., Visual Event Perception. In: Handbook of Sensory Physiology. R. Held, H.W. Leibowitz, and H.L. Teuber (eds.) Springer Verlag, Berlin, Heidelberg, Vol. III, 1978b.

Johansson, G., About perspective transformations and the theory of visual space perception. Department of Psychology, University of Uppsala, Sweden. Uppsala Psychological Reports, No. 278, 1980.

Johansson, G., von Hofsten, C. and Jansson, G., Event Perception. Annual Review of Psychol., 1980, 31, 27-63.

Land, E.H., and McCann, J.J., Lightness and the Retinex theory. J. Opt. Soc. Am., 1971, 61, 1-11.

Lee, D.N., Visual information during locomotion. In: Perception: Essays in Honour of James Gibson. R.B. MacLeod and H.L. Pick (eds.), Cornell Univ. Press, Ithaca/London, 1974, 250-267.

Lee, D.L. and Aronsson, E., Visual proprioceptive control of standing in human infants. Perception and Psychophysics, 1974, 15, 3, 539-532.

Nakayama, K., and Loomis, J.M., Optical velocity patterns, velocity-
    sensitive neurons and space perception: a hypothesis.
    Perception, 1974, 3, 63-80.
Prazdny, K., Egomotion and relative depth from optical flow.
    Biol. Cybernetics, 1980, 36, 87-192.
Ullman, S., The interpretation of visual motion. MIT Press,
    Cambridge, Mass., 1978.
Wallach, H., On Perception. Quadrangle/New York Times Book Co.,
    1976.

THRESHOLDS OF MOTION PERCEPTION

Claude Bonnet

Lab. de Psychologie Expérimentale
associé au C.N.R.S.
75006 Paris, France

INTRODUCTION

Threshold measurement is the most fundamental approach in the study of the sensitivity of living organisms to motion. Such studies started more than a century ago. The variety of experimental situations is still increasing. It is believed that an attempt for a single conceptual background is necessary in order to elucidate the processes at work in motion detection.

By definition a motion threshold is the limit for the detection of *the direction of the translation of a moving target* expressed in terms of one stimulus dimension. There are two such limits which define the visual motion domain. The upper limit, sometimes called the fusion threshold, has received little attention. Exceptions are researches made by Brown (1958), Kaufman et al. (1971), Caelli et al. (1978) and van der Glas et al. (1979). The present review will therefore concentrate on the lower limit.

Each of the variables which influence the motion threshold can be considered as an axis of the multidimensional space within which motion sensitivity is defined. However, two dimensions are basic. One is spatial, the other one is temporal. Traditionally, motion stimuli were defined in an amplitude-time domain such as the spatial extent and the duration of a translation. More recently, motion stimuli are described in a frequency domain, i.e. in terms of their Spatial and Temporal Frequency content. As it will appear in the following, each of these two cases of description is more relevant to a certain kind of motion display, and also to a certain type of process model. A description in the amplitude-time domain is generally used with models in which the visual system is considered as a set of trigger-features. A description in the frequency domain is generally used with models considering the visual system as made of frequency filters.

In this paper an integrative point of view will be followed.
Emphasis will be placed on the complementarity of the information
given by each approach.

1. A CLASSIFICATION OF MOTION DISPLAYS

Three classes of motion displays are used in the measurement
of thresholds for targets translating in a fronto-parallel plane.
They will be termed Single motion displays, Oscillatory motion displays and Frequency motion displays. Each class of displays is
defined by the variables which describe the stimulus. These variables also define the domain in which the results are directly expressed. From a theoretical point of view, only two domains are
necessary: a space-time domain and a frequency domain. The present
taxonomy uses three classes for empirical reasons. Oscillatory
motion displays (§ 1.2) can then be viewed as a linking class between Single displays (§ 1.1) and Frequency displays (§ 1.3).

Within each class of displays different variables are used in
order to estimate motion sensitivity. Only the procedures met in
the literature have been summarized.

1.1. Single motion

A single motion display is one in which at every trial a single
translatory motion of a single target is presented. The *amplitude*,
i.e. the angular extent, the *exposure time* and consequently the
velocity of the motion are specified. The velocity function generating the motion with such a display is a linear function in the Amplitude-Time domain i.e. a constant velocity (see Figure 1).

Four procedures have been used in motion threshold measurement.
In a constant duration procedure, the exposure time remains constant
from trial to trial while velocity and amplitude co-vary. In a constant amplitude procedure, the amplitude remains constant while exposure time and velocity counter-vary. The thresholds thus obtained
are called velocity thresholds by Graham (1965). In a constant velocity procedure, the velocity remains constant while exposure time
and amplitude co-vary. This procedure yields what Graham (1965) has
called displacement thresholds. No specific effect of the procedure
is expected (Bonnet, 1975). Data obtained with one procedure are
predicted to be identical to data obtained with the other procedures.
This is suggested by the results of Graham (1968, p. 25) and of Cohen
and Bonnet (1972). The fourth procedure uses changes in the *luminance*
of the target (Pollock, 1953; Barbur and Ruddock, 1980). The results
can then be expressed in terms of (luminance) contrast:

$$C = \frac{L_T - L_B}{L_B}$$

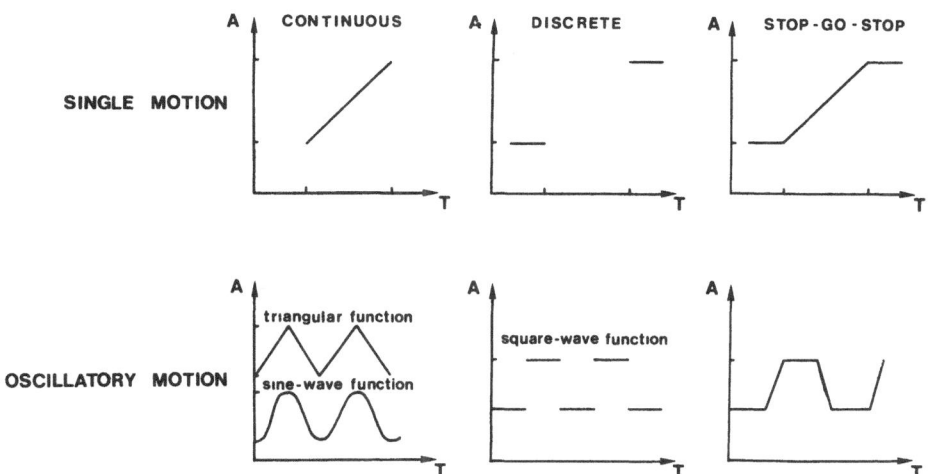

Fig. 1. Motion displays in an Amplitude-Time space. Frequency motions
have not been represented.

where C is contrast, L_T the luminance of the target and L_B the luminance of the background.

With such a Single motion display, three modes of presentation can be distinguished:

1.1.1. Continuous Single Motion

The target is continuously in motion in a single direction during the exposure time of its translation which contains no stationary component. See also § 1.3.1.

1.1.2. Discrete Single Motion

The translation of a single target is produced by its successive appearance in successive but spatially separated positions. Traditionally, only two positions are used as in most studies of apparent (beta) motion (Exner, 1875; Wertheimer, 1912; Korte, 1915; Neff, 1936; Aarons, 1964). A Discrete Single Motion is described physically by the durations of the stationary periods, of the Inter-Stimulus Interval (ISI) and by the spatial separation of the stationary appearances (A). Stimuli appearing repeatedly at successive positions as in the display used by Kolers (1972) and by Morgan (1979) also belong to this class. Obviously, when the stationary periods of each stimulus appearance, the Inter-Stimulus time Interval (ISI) and spatial separation become short enough (as they could be with a computer driven CRT) such a display may be shown to become functionally equivalent to a Continuous Single Motion display. Systematic experiments remain to be done in order to study the transition of the psychophysical functions from Discrete-like to

Continuous-like data.

Phenomenally, a Discrete Single Motion may raise different experiences such as the appearance of a unique target, the simultaneous appearance of two targets, an apparent or stroboscopic motion (Wertheimer's beta motion), or finally the successive appearance of the same object in two separate places. The apparent motion phenomenon will not be considered in the present paper (see reviews in Kolers, 1972; Anstis, 1978).

1.1.3. Stop-go-stop Single Motion

Combining a Continuous and a Discrete Motion provides a motion display which contains both a real motion phase and two stationary components.

1.2. Oscillatory Motion

While single motion is always unidirectional during a given trial, an oscillatory motion consists of a repetitive alternation of two translations in opposite directions. Such an oscillatory motion is physically described by its Amplitude (A), and its Temporal Frequency (TF). Given a particular amplitude and temporal frequency, the velocity is determined. For a given velocity function, different procedures can be used for measuring motion thresholds. In a *constant amplitude procedure* the amplitude of oscillation remains constant and motion thresholds are measured by a change in Temporal Frequency. In a *constant frequency procedure* the temporal frequency of oscillation remains constant and motion thresholds are measured by a change in the amplitude of oscillation. In a third procedure Amplitude and Temporal Frequency remain constant and motion thresholds are measured through changes in the *luminance* of either the target or the background and can be expressed in terms of contrast as in the case of a Single Motion.

The classification (see Fig. 1) in three modes of presentation which have been proposed for continuous motion holds also for oscillatory motion.

1.2.1. Continuous oscillatory motion

Three velocity functions generate continuous oscillatory motion. In a Sinusoidal velocity function, also called harmonic motion, the velocity changes as a sinusoidal function of time. In a triangular velocity function, the target moves to and fro between two extreme positions at a constant velocity. In a saw-tooth velocity function the target moves at a constant velocity, returns to its origin at an infinitely high velocity and then starts again. Such motions have been used by Krauskopf (1957), Tyler and Torres (1972), King-Smith (1978a, b).

1.2.2. Discrete oscillatory motion

A discrete oscillatory motion is obtained in using a square-

wave velocity function. In such a mode of presentation the target
is seen stationary at two alternating positions separated in space,
but the Inter Stimulus Time Interval is nil. The same percepts as
those described for discrete single motion are reported. Another
apparent motion phenomenon should be mentioned here: the omega movement described as a shadow oscillating between two lights (Saucer,
1953, 1954).

1.2.3. Stop-go-stop oscillatory motion
Such a mode of presentation has been used for instance by
Kaufman et al. (1971) in the measurement of the upper threshold
for motion.

1.3. Frequency Motion

In this class of motion displays, a repetitive pattern is generally drifting in a single direction. Such translations of the
pattern are described in a Spatial Frequency - Temporal Frequency
domain. Spatial Frequency (SF) is expressed in terms of number of
cycles of the repetitive pattern per degree of visual angle (cpd).
Temporal Frequency (TF) is expressed in terms of number of cycles
per second (or Hz). The velocity of a unidirectional drift is given
by the ratio of the Temporal over the Spatial Frequencies.

1.3.1. Frequency motion generally treated as Single Motion
Brown (1931) and Leibowitz (1955a) have used a train of circular or square targets moving along a rectilinear track. Each black
target is separated from the next one by a large interval so that
the luminance profile along the direction of drift has a rectangular
shape. Presumably, the greater the target interval cycle of such a
repetitive pattern, the more it gives results identical to a single
stimulus (see Fig. 1 in Bonnet 1975). Actually, such stimuli
have traditionally been treated just as a Single Motion, the results
being expressed in the Amplitude (or Velocity)-Time domain. The
justification for such a treatment is that these stimuli have a
rather complex description in the spatial and temporal frequency
domain.

1.3.2. Drifting gratings
The most representative instance of a frequency motion display
is given by drifting sinusoidal gratings. The luminance modulation
of stimuli in single or oscillatory motion, as discussed previously,
has a rectangular profile. This third dimension has been omitted
from their representation in Fig. 1. A sinusoidal vertical grating
drifting rightward has a luminance profile defined (Sekuler and
Levinson, 1974) by:

$$L (x , t) = L_o . (1 + M. \cos (fx \pm wt))$$

where L (x , t) is the luminance at the position x at time t, L_o the mean luminance, $f/(2\pi)$ the spatial frequency in cpd and $w/(2\pi)$ the temporal frequency in Hz; M is the modulation or Michelson contrast given by:

$$M = \frac{L_{MAX} - L_{MIN}}{L_{MAX} + L_{MIN}}$$

where L_{MAX} is the maximum luminance of the grating and L_{MIN} the minimum luminance. The term modulation is to be preferred to that of contrast which should be reserved for luminance contrast as defined above (§ 1).

Motion thresholds are generally estimated through changes in the modulation of the gratings. Strictly speaking, *motion thresholds should designate only those measurements in which a judgment of direction of motion is required*. This issue is relevant since changing instructions not only changes the level of the estimated motion sensitivity, but also the shape of the sensitivity function (see § 2.2.).

There is one instance in which the liminal Velocity for drifting gratings has been measured (van der Glas, et al., 1979).

1.3.3. Temporal shift of Phase

Instead of presenting waves travelling in a constant direction, gratings can be presented with a contrast reversal or a counterphase procedure.

In contrast reversal, the phase of the grating is varied as a square-wave function of the time by an amount of 180° so that the lighter parts become darker and the darker parts become lighter. In a counterphase procedure, the phase is continuously changing bidirectionally. These two procedures can be described trigonometrically as the sum of two motions of opposite directions (Sekuler & Levinson, 1974; Kulikowski, 1978). The luminance profile of a counterphase grating is defined by:

L (x , t) = L_o . (1 + 0.5 M. cos (fx - wt) + 0.5 M. cos (fx + wt)).

2. ASSUMPTIONS AND PROCESSES IN MOTION DETECTION

2.1. Variables for thresholds

The lower thresholds for directional motion detection have been measured in terms of Velocity (V_{lim}), Amplitude (A_{lim}), Duration (T_{lim}), Temporal Frequency (TF_{lim}), Contrast of Luminance (C_{lim}) or Modulation (M_{lim}). Other variables can also be used such as Spatial Frequency. The choice of the dependent variable in which the threshold is expressed is mostly a matter of convenience. The strategy of threshold measurement consists in defining the zero point of motion sensation in terms of the stimulus dimensions. The lower the value (intensity or extent) of a given stimulus dimension necessary to reach the criterion of detection of the direction of motion, the greater the sensitivity of the organism. Motion sensitivity requires a multidimensional space to be defined. For simplicity, results obtained with a Frequency Motion display are generally defined in a three dimensional space: Modulation - Spatial frequency - Temporal frequency (see for instance Kelly 1977, 1979). In this instance at least one fourth dimension is omitted such as the mean retinal illuminance level. Another important dimension to be specified is the retinal eccentricity since the eye is not an homogeneous receptor (see § 4).

Hence, any representation of motion sensitivity will only be partial. Bearing this point in mind, there are two main domains in which motion sensitivity can be represented: the amplitude-time domain and the spatial frequency - temporal frequency domain. The experimental variables of a Single Motion are the angular extent and the duration of the translation. Results will then be expressed in the Amplitude-Time domain. A Frequency motion of a sinusoidal grating is defined by the Spatial Frequency of the grating and the Temporal Frequency of its drift. Motion sensitivity in terms of modulation will then be expressed in the frequency domain. For more complex stimuli such as a random pattern, the choice of either one or the other domain is mostly a matter of convenience. Mathematically it is possible to transform data expressed in one domain into the other.

Aside from their mathematical equivalence, the use of either one or the other of these two domains stresses different but complementary aspects of the ways higher mammals process motion information.

It has been said above for some motion displays that they could be classified in one class or in another one. The decisive criterion should be neither the phenomenal appearance of the motion nor the physical conditions by themselves but the psychophysical effect of these conditions with respect to the assumed underlying process.

2.2. Motion detection: two analysing systems

Motion threshold data cannot be reviewed without taking into consideration the processes assumed to be at work. Almost every proposed model comes to assume two different ways of processing visual motion information. Actually all of the proposed models, which range from conceptual to semi-quantitative models, fit into two related classes: the *trigger-feature models* or the *frequency filter models* (Bonnet, 1977).

Exner (1875, 1888) assumed that for a slow velocity translation, motion is inferred from successive positions. For medium to high velocities, motion would be perceived directly. Exner tried to imagine the neural circuitry upon which such direct perception of motion could be based. Such a two process assumption was renewed by Leibowitz (1955a, b) and Brown (1955). They proposed that for high velocities and/or for short exposure times, motion would be detected directly and mediated by a "single sensory event". Along that same line of reasoning Bonnet (1971, 1975, 1977) has tried to develop a conceptual model which assumes that visual motion information can be processed by either one of two functionally separate systems: the *Movingness Analysing System** (MAS) or the *Displacement Analysing System* (DAS). The empirical evidence for such a model was initially based on the analysis of Single (or Single-like) motion configurations. Such a model was described as a two feature analysing system (Bonnet, 1977).

The Displacement Analysing System (DAS) would process the spatial and stationary components of any motion or in other words the *amplitude* of any translation. Kinchla and Allan (1969) and Kinchla (1971) have provided speculations about the basic process for such a system. The information concerning the amplitude would be obtained by comparing the actual position of the target (ending position) with the sensory trace of a previous position.

The Movingness Analysing System (MAS) would process the motion information as such (i.e. velocity).

The main experimental evidence for such a model consists in showing how in some cases in which temporal summation can be demonstrated (see § 3.2.) stationary marks are ineffective. Reciprocally when stationary marks are effective, temporal summation fails to appear. In this respect, the three modes of presentation of a single motion are paradigmatic (Bonnet, 1975, 1977).

On the basis of Oscillatory motion displays, Tyler and Torres (1972) and King-Smith (1975a, b) provide arguments in favor of the assumption according to which the organism can process predominantly either velocity-features (MAS) or amplitude features (DAS) (see Bonnet, 1977). In that respect, the three possible modes of presentation of the Oscillatory motion are also paradigmatic as in the case of Single motion.

The second main tradition of theorizing about motion processing refers to the filtering characteristics in the spatial frequency and/or temporal frequency domains. A *sustained* (or pattern) mechanism (or channel) would process spatial information while a *transient* (flicker or movement) mechanism would process temporal (or movement) information (Keesey, 1972. Kulikowski and Tolhurst, 1973. Tolhurst, 1973). For a low SF grating drifting at a middle Temporal Frequency, the modulation contrast of which is increasing, the first threshold would correspond to a flicker experience or to a directional motion experience. A greater modulation contrast will be necessary for the subject to be able to detect the spatial structure of the pattern: its orientation for instance. For a high

---

* The term Movingness is preferred to the term motion since the latter is currently defined as a change in place with time, i.e. a displacement.

SF grating the spatial structure will be detected with a lower modulation contrast than the direction of motion. In a complementary manner, for a middle range Spatial Frequency with a low Temporal Frequency drift, the spatial structure will be detected at a lower modulation contrast than the direction of motion. Conversely at a high TF drift the direction of motion will be detected at a lower modulation contrast than the spatial structure (see Kulikowski and Tolhurst, 1973. Sekuler et al., 1978). There is some confusion in the literature between flicker and motion experiences the former being actually less reliably reported than the latter (Gorea, 1979; Richards, 1971).

For instance Richards (1971), using a rotating Exner-spiral, as a motion display, showed that the modulation threshold for detecting this kind of motion first declines with the increase in velocity, then rises with a further increase of its velocity. In the range of high velocities, at a first modulation some non-directional temporal modulation, which Richards calls scintillation, is detected. Higher modulation is needed before a directional motion is going to be detected.

In the following, these models will be described in terms of their processing differences. A constant response criterion of *directional motion* is necessarily assumed. For that reason the two systems should not be called motion (or flicker) vs. pattern but preferentially Movingness vs. Displacement Systems. In effect, the characteristics of the MAS and of the DAS defined in the space-time domain (Bonnet, 1975) should be consistent with their characteristics when defined in the frequency domain (Bonnet, 1977, King-Smith, 1978a, b). At present such a statement cannot be quantitatively proved because of the very large and somehow uncontrolled differences in the experimental situations. But its conceptual consistency should encourage further attempts in that direction. To sum up, the Movingness Analysing System would be made of a band-pass Temporal Filter and a low pass Spatial filter, while the Displacement Analysing System would be made of a band-pass Spatial Filter and a low pass Temporal Filter.

3. EXPERIMENTAL FINDINGS

3.1. Spatio-Temporal sensitivity for motion detection

As is now well known, the sensitivity function for stationary sine-wave gratings is typically a band-pass function: sensitivity first increases as the square of the SF in the low range, reaches an optimum between 2 to 5 cpd (in photopic vision) and then declines exponentially with SF in the high range (Campbell and Green, 1965; Kelly, 1977). Such a sensitivity function is used to define a spatial filter. In the spatial domain, it has a Mexican hat shape as the one used in King-Smith's model (1978a, b).

We will examine first (§ 3.1.1. and 3.1.2.) the effects of

temporal modulation on (pattern) sensitivity before analysing motion
sensitivity as such.

3.1.1. Flashing standing sine-waves

A temporal modulation of a stationary grating introduces changes
in the sensitivity function. When the exposure time is longer than
some 1000 msec, flashing standing sine-waves typically result in no
change in sensitivity in the high SF range (Kelly, 1977; Tulunay-
Keesey and Jones, 1976). For shorter exposure times, there is a
drop of sensitivity in the high SF range: for a given high SF (for
instance 10 cpd) sensitivity increases monotonically with exposure-
time up to a critical duration (Kelly, 1977; Tulunay-Keesey and
Jones, 1976; Breitmeyer and Ganz, 1977; Legge, 1978). In the low SF
range (for instance 0.5 cpd) decreasing exposure time first improves
sensitivity (Kelly, 1977, his Fig. 4 and 5). Then a further decrease
in exposure time drops sensitivity at any SF: the sensitivity function becomes a low-pass function and sensitivity appears to be in-
versely proportional to the energy of the stimulus. As a matter of
fact a trade-off has been demonstrated between modulation and exposure time of standing sine-waves (Arend, 1976; Breitmeyer and
Ganz, 1977; Gorea, 1978; Arend and Lange, 1979). There is some indication that the critical duration (τ) beyond which such a trade-
off holds increases with SF of the gratings.

3.1.2. Flickering standing sine-waves

Similar conclusions about temporal summation can be obtained
with flickering standing sine-waves. The results will now be expressed in the Temporal frequency domain. Using counterphase modulations, Robson (1966), Kelly (1972), Kulikowski and Tolhurst (1973)
and Levinson and Sekuler (1973) showed that, typically, the sensitivity as a function of spatial frequency is hardly affected by
temporal modulation below some critical temporal frequency ($\frac{1}{2}\tau$). This
holds true in the high SF range. However, temporal modulation
improves sensitivity in the low SF range. Beyong a critical Temporal Frequency ($\frac{1}{2}\tau$), the sensitivity function becomes a low-pass
function, just as in the case of a single flash exposure time. A
trade-off is observed between sensitivity and time period.

When expressed as a function of Temporal Frequency with Spatial
Frequency as a parameter, the same data demonstrated that sensitivity is a band-pass function of temporal frequency for a low SF (for
instance 0.5 cpd). There is then a unique TF for which sensitivity
is maximal. An increase in Spatial Frequency (for instance 4 cpd)
increases sensitivity for low TF and hardly affects it for higher
TF. With a further increase in SF (16 or 22 cpd) sensitivity is
clearly of a low-pass type. For any TF, sensitivity is now lower
the higher the Spatial Frequency. A spatio-temporal threshold surface (in the frequency domain) can be constructed which shows the
combined effects of spatial and temporal modulations on sensitivity. The question remains largely open if such a surface is uni-
modal (Kelly, 1977) or bimodal as the assumption of sustained vs.

transient mechanisms would suggest (Legge, 1978 for instance).

3.1.3 Traveling sine-waves

Now the question is to know if motion considered as a spatio-temporal modulation including an actual directional component (Sekuler, 1975) has a genuine effect upon visual sensitivity. Clearly, in natural conditions eye movements (tremor, drifts or even micro saccades*) introduce some movements in the retinal image of a stationary stimulus which would otherwise vanish rapidly. Such eye movements influence contrast sensitivity to spatio-temporal patterns (Kulikowski, 1971; Tulunay-Keesey, 1978). Levinson and Sekuler (1975) showed that sensitivity to drifting gratings is consistently better than sensitivity to flickering gratings by a factor of 2 in a restricted range of Temporal and Spatial Frequencies. Kelly (1979), with a stabilized vision technique, confirmed that the factor of 2 holds for the ratio of sensitivities to drifting and flickering sine gratings for SF between 0.37 cpd to about 15 cpd and TF between 1 Hz to 30 Hz. In natural vision, the difference may be some what reduced or even suppressed in the high SF range (Tolhurst, 1973; Kulikowski and Tolhurst, 1973).

Now two questions remain open in the analysis of sensitivity to travelling waves which are of major interest for a model of motion detection. The first question is about a Spatial Frequency vs. Velocity trade-off and the second and related question is about the unimodality or bimodality of the spatio-temporal threshold surface.

3.1.4. A Spatial Frequency x Velocity trade-off?

Using square-wave drifting gratings, the fundamental SF of which ranged between 1.25 and 4.28 cpd, Crook (1973) measured the luminance threshold for the detection of the direction of motion at different velocity drifts. The best or optimal sensitivities to motion corresponded to a unique Temporal Frequency, suggesting a SF vs. Velocity trade-off. Watanabe et al. (1968) measured the Modulation thresholds for a range of Spatial Frequencies of sine-wave drifting gratings between 0.087 cpd and 4.35 cpd with velocities ranging between about 1 to 80°/sec. When sensitivities are expressed as a function of Spatial Frequency, with velocity as a parameter, there are as many sensitivity functions as velocities. When the same results are expressed as a function of velocity with Spatial Frequency as a parameter, every SF corresponds to a different Velocity optimum. In the low Spatial Frequency region, the optimum velocity is a function of the Spatial Frequency. Now, if the sensitivities were plotted as a function of Temporal Frequency with SF as a para-

---

* A tremor is a relatively high frequency and low amplitude eye movement. A drift is a slow velocity eye movement. A micro saccade is a fast velocity movement of the eye over a small amplitude.

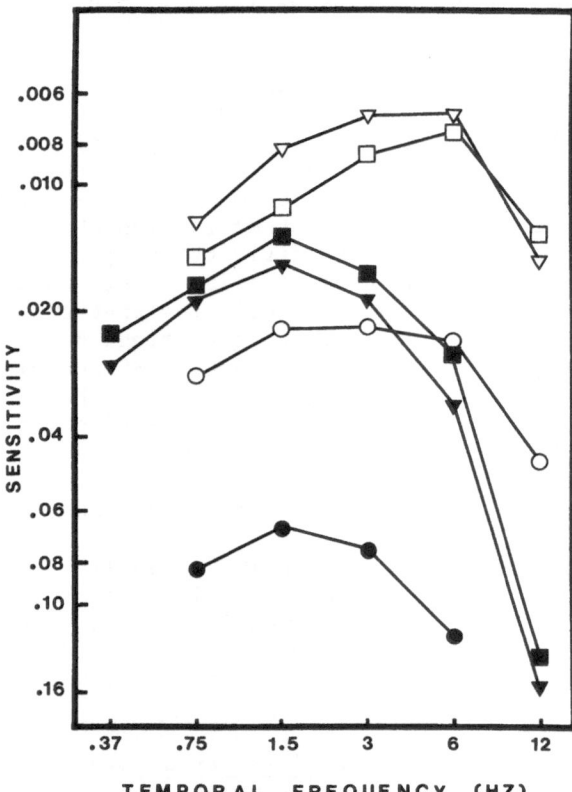

Fig. 2. Sensitivity to motion as a function of the Temporal Frequency
        of drift of sine-wave gratings. Open symbols for 3 cd.m^{-2};
        dark symbols for .15 cd.m^{-2}; square = .84 cpd; triangle
        3.4 cpd and disk = 8 cpd.

meter, every curve shows an optimal sensitivity for TF of about 6 Hz
while there are differences in level due to SF. Similar conclusions
are obtained by Tolhurst et al. (1973) and Sharpe (1974). Actually,
it is clear that such a SF x Velocity trade-off would only hold when
Spatial Frequency is in the low range (Sekuler et al., 1978). A decrease in the retinal luminance shifts the optimal TF of the sensitivity function towards a lower TF (van Nes et al., 1967) as it
shifts the optimal SF of the sensitivity function towards a lower SF
(Campbell and Robson, 1968). Hence a decrease in retinal luminance
should shift the optimal TF for which the SF x Velocity trade-off
holds towards a lower value. High luminance levels (L = 100 cd.m^{-2})
have been used by Tolhurst et al. (1973) who reported an optimum of
6 Hz. Figure 2 reports the results of an unpublished experiment in

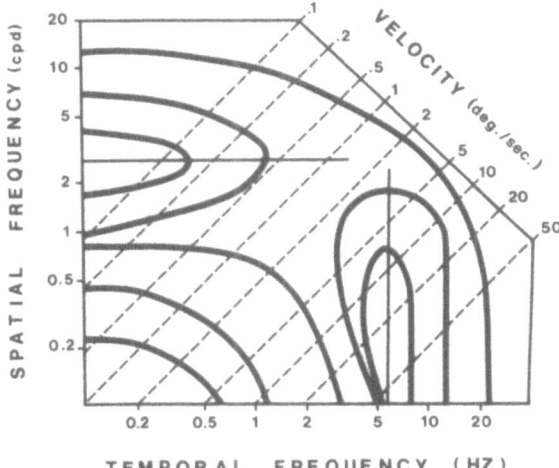

Fig. 3. Diagrammatic representation of isosensitivity curves for
       motion detection illustrating the bimodality case (see text).

which modulation thresholds for detection of the direction of motion
have been obtained at 3 cd.m^{-2} and at .15 cd.m^{-2}. The SF x Velocity
trade-off holds for the two lower Spatial Frequencies (.84 cpd and
3.4 cpd). For the higher SF (8 cpd) no such trade-off is shown. The
optimal frequency is higher (6 Hz) for the higher luminance and much
lower (1.5 Hz) for the lower mean luminance level. Such a SF x Velocity trade-off can also be inferred from other authors' results as
in Kelly (1979) or Koenderink and van Doorn (1979).
      The optimal Temporal Frequency of the trade-off is the counter-
part in the frequency domain of the critical duration (τ) for temporal integration (Ganz, 1975):

$$\tau = P_{opt}/\ 2$$

where P_{opt} is the temporal period corresponding to the optimal Temporal Frequency. Hence temporal summation should appear in the high
TF range and not in the low TF range. On a plot of isosensitivity
curves in the spatio-temporal plane such as those drawn by van Nes
(1968), Kelly (1979) or Koenderink and van Doorn (1979), isosensitivity curves would appear parallel to the Spatial Frequency axis when
the trade-off holds and would shift from that orientation when the
trade-off fails. This is sketchily reconstructed in Figure 3. At
the moment such a representation is a speculative way to look at a
set of pooled results. In effect, in none of the three studies have
the authors tried to validate such an assumption. In Kelly's data
(his Fig. 15, p. 1347), the maximum sensitivity at every velocity
which would verify the trade-off assumption seems to hold for a

range of SF below 0.7 cpd. Although one may argue about such a conclusion, it is nevertheless obvious that in the three sets of results, below some optimal TF of about 7 Hz, isosensitivity curves
are less and less dependent upon TF. Actually, they become dependent
only upon Spatial Frequency when SF becomes greater than a certain value.

Now if these data are going to be used as evidence for a
two-process model, rather than a continuum (cf. Kelly, 1979) some
kind of bimodality must be demonstrated. Again, the evidence of bimodality in the spatio-temporal frequency surface has not been
looked for systematically. The only clear evidence is given by
Koenderink and van Doorn (1979): for instance sensitivity as a function of TF declines as a function of the increase of TF for a drifting grating of 4 cpd, while for a SF of .25 cpd the function varies
curvilinearly. The sensitivity level at which the two curves inter-
sect is lower than the optimum sensitivity for each SF. A similar
bimodality is observed in the Sensitivity-Spatial Frequency plot.
Isosensitivity curves in the spatiotemporal frequency plane, as
given by these authors, summarize such a bimodality with a first
maximum for a grating of about 5 cpd drifting at 0.1 Hz, and a
second maximum for a grating at about 0.2 cpd drifting at 7 Hz.
Koenderink and van Doorn (1979) reanalysing some previously published results concluded that examination of the data of Robson (1966)
and van Nes (1968) allows for the conclusion of bimodality, which is
not stressed by these authors. No such conclusion holds for Kelly's
data.

Between the above reported studies, there are differences in
the criterion used for estimating the modulation threshold. Some
authors asked their subject to detect any heterogenity in the stimulus: whether it is spatial, temporal or directional. Some other
authors asked their subjects to use one among two criteria as separate occasions: a pattern criterion or a motion (or flicker) criterion. With regard to the present question, only the results obtained
when the subject is asked to detect the direction of motion are
directly conclusive. On that basis, the main clear separation between
the Movingness Analysing System and the Displacement Analysing System
is the form of the motion sensitivity function.

In the MAS, the motion sensitivity is a low-pass function of
Spatial Frequency and a band-pass function of Temporal Frequencies.
It is prominent when low SF gratings drift at medium TF. The exis-
tence of a Spatial Frequency x Velocity summation characterizes the
MAS. In the DAS, the motion sensitivity is a band-pass function of
Spatial Frequency and a low-pass function of Temporal Frequency. It
is prominent when medium to high SF gratings drift at low TF.

Hence it is suggested that the best indication of MAS prominence
will be found in summation relationships while DAS prominence will
be demonstrated by the importance of the spatial content of the
stimulus. The stationary components of the stimulus will, in this
respect, deserve particular attention.

3.2. Processing velocity vs. Processing amplitude

Comparison of continuous motion with discrete motion can provide insights about whether the organism is, in a given situation,
processing the velocity contained in the motion phase, or the amplitude contained in the stationary components of the motion display.
Stop-go-stop presentation could be used to reinforce the conclusion,
depending upon which of the two other modes raise more similar results (Bonnet, 1975, 1977).

With Continuous Single Motion, it has been repeatedly demonstrated that when exposure time is below some critical duration (τ)
a velocity x time trade-off is obtained (Dimmick and Karl, 1930;
Leibowitz, 1955a; Cohen and Bonnet, 1972; Johnson and Leibowitz,
1976). Beyond this critical duration (τ), the trade-off does not
hold, and the motion threshold corresponds to a constant and low
velocity, irrespective of the exposure time. Below τ the velocity
vs. time trade-off means that irrespective of the exposure time, a
motion threshold corresponds to a constant and low amplitude of the
continuous motion.

Although thresholds are similar beyond the critical τ duration
of exposure time, or ISI, whether they have been obtained with
Continuous, Discrete or Stop-go-Stop modes of presentation (Bonnet,
1975), the two latter modes generate lower Amplitude thresholds than
the first one in the low exposure time range. As a matter of fact
within a large range of ISI's the liminal Amplitude for detecting a
directional component in a Discrete Single Motion increases with ISI
(Neuhaus, 1930; Sgro, 1963; Kahneman, 1967; Kahneman and Wolman,
1970; Kolers, 1972). The three modes of presentation were used in
a replication of Bonnet's experiment (1975). For Discrete and Stop-
go-Stop conditions each stationary period lasted 10 msec. In this
replication the mean luminance of a 2.4 min of arc spot was maintained constant at 3.2 cd.m^{-2}. The threshold criterion was detecting the directional component of the translation. Other specifica-
tions are those of a previous experiment (Bonnet, 1975). As shown
in Figure 4 for results expressed in liminal Amplitude (A_{lim}), a
V_{lim} x T trade-off is obtained below an exposure time of 180 msec
only for Continuous Single motion. Stop-go-Stop and Discrete modes
showed similar thresholds. In both conditions, A_{lim} increases with
exposure time. These thresholds are lower than those given by the
Continuous mode when T $\leq$ 180 msec. Beyond that duration the Continuous mode shows thresholds similar to those obtained in Discrete
and Stop-go-Stop modes. Expressed in different terms, for values
greater than τ the liminal velocity for motion detection is nearly
constant and slow in the three modes.

An apparently paradoxical feature should be mentioned: beyond
τ, the Discrete mode generates the perception of a discrete jump of
the target between two positions, while apparent beta motion is
only reported for an ISI $\leq$ τ. Hence a phenomenal resemblance cannot
explain the similarity of the processes.

The proposed interpretation of the present results implies that

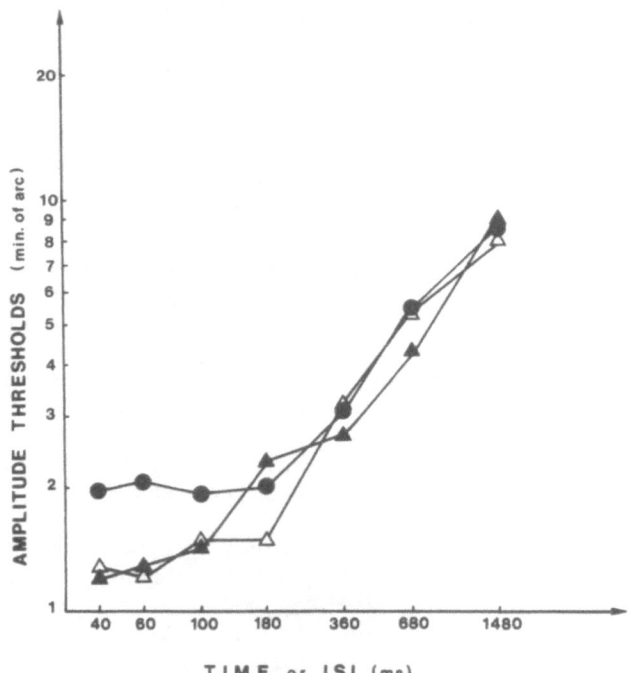

Fig. 4. Liminal Amplitude for motion detection as a function of
exposure time on log-log coordinates for Continuous (● - ●),
Stop-go-Stop (▲ - ▲) and Discrete (Δ - Δ) Single Motion
displays. Stationary periods are 10 msec.

when stationary components are present (as in Discrete and Stop-go-
Stop modes), motion is detected on the basis of its amplitude, while
when the display contains only a motion phase (as in the Continous
mode motion is detected on the basis of its velocity. In the latter
case, when velocity is low enough (equivalent to when the exposure
time is long enough), the Continous mode behaves as if it contained
stationary components, i.e. it seems to be detected on the basis of
its amplitude. As mentioned earlier (Bonnet, 1975, 1977), the stationary periods of the Stop-go-Stop mode appear to be effective only
when their durations are short. Clearly, with long and unspecified
stationary periods, the Stop-go-Stop mode shows thresholds similar
to those of the Continuous mode, including a Velocity x Time trade-
off (Leibowitz, 1955a; Johnson and Leibowitz, 1976).
      Similar conclusions can be reached from the analysis of Oscil-
latory motion displays. Krauskopf (1957) used a bright line oscillating sinusoidaliy with a maximum peak-to-peak amplitude of 4 min
(Continuous mode) within a stabilized viewing condition. He compared
luminance Contrast thresholds for an oscillating line to the con-

trast threshold for detecting a stationary line. Sensitivity for
a moving line is better than that for a stationary line when the
Temporal Frequency is greater than 10 Hz. King-Smith and Riggs (1978)
and King-Smith (1978a, b) used a similar display. The authors
compared sensitivity for a triangular motion (Continuous mode) to
sensitivity for a square-wave motion (Discrete mode, with ISI = 0)
when both oscillating at the same frequency (1 Hz). For both modes,
sensitivity varies curvilinearly with amplitude. The best sensitivity
for the triangular motion is obtained for a peak-to-peak amplitude
of 32 min of arc. For the square wave motion, the best sensitivity
occured at 8 min. of arc. In the low range of amplitudes (< 8 min)
sensitivity is better for Discrete Oscillations than for Continuous
ones. A reverse tendency is observed for larger amplitudes. At its
optimal amplitude, sensitivity for Discrete oscillation is 1.5 times
lower than the best sensitivity for Continuous oscillation.

Contrast thresholds for detecting the presence or for detecting
the motion of the line are similar: over 1 min amplitude of oscillation for the Continuous motion and over 0.5 min amplitude of oscillation for the Discrete motion.

Now, when the Temporal Frequency of oscillation of a sinusoidal
Continuous motion is varied over a large range, maintaining a constant and small amplitude (3 min), sensitivity appears nearly independent of the frequency below 8 Hz (King-Smith et al., 1977). With
a triangular Continuous motion, when the amplitude varies in inverse
proportion to the Temporal Frequency, the Velocity is kept constant.
In such circumstances, King-Smith and Riggs (1978) showed that
motion sensitivity is independent from TF when velocity is 2.13°/sec.
In other words, an Amplitude x Temporal Frequency trade-off is observed which is equivalent to the Velocity x Time trade-off obtained
with Single motion. Complementarily, with a lower velocity (16 min/
sec) sensitivity declines when TF increases. Hence with such a low
velocity no trade-off is obtained.

In conclusion of these researches, King-Smith (1978a, b) stated
that at low velocities and small amplitudes, motion is detected on
the basis of the amplitude, while at higher velocities and larger
amplitudes it would be detected on the basis of velocity.

Complementary information is given by Frequency motion displays. Levinson and Sekuler (1975) have compared motion sensitivity
for drifting gratings and for counterphase gratings (i.e. sinusoi-
dally modulated in time). For low Spatial Frequencies (< 7 cpd) the
modulation thresholds for drifting gratings (4 < TF < 8 Hz) was
half the threshold for counterphase. In other words the threshold
is equal to the contrast of one of the directional components of
the counterphase grating. Such an analysis led Levinson and Sekuler
(1975) to conclude that the two directional components of the
counterphase grating can be detected independently.

Kulikowski (1978) verified these results for the same range of
spatial and temporal frequencies. He obtained a different result
for SF above 20 cpd. Sensitivity for a contrast reversal (square
wave function) was measured with two phenomenal criteria (pattern

vs. movement) as a function of the Spatial Frequency at TF = 1.67 Hz.
Kulikowski (1978) showed that sensitivity for detecting motion declines faster than sensitivity for detecting the pattern. In a
further condition motion sensitivity for a grating drifting at a
rate of 1.67 Hz was measured. A better motion sensitivity was found
for drifting gratings than for contrast reversals. A factor of $4/\pi$
separated both results in the lowest Spatial Frequency range. Beyond
a SF of about 20 cpd motion sensitivity for the drifting gratings
became suddenly much closer to sensitivity for detecting the pattern
in a contrast reversal condition. The same results hold at Temporal
Frequencies of 2 and 5 Hz. For higher Temporal Frequencies (> 10 Hz),
sensitivity for drifting gratings remains always comparable to sensitivity for contrast reversal with a movement criterion.

3.3. Further considerations

3.3.1. Processing stationary positions

The optimal beta apparent motion, found when using the Single
Discrete mode, shows a reciprocity between Amplitude and ISI. This
was predicted by Korte's third law (Korte, 1915; Neuhaus, 1930;
Kinchla and Allan, 1969; Caelli et al., 1978, see also figure 4).
With an Oscillatory Discrete mode (square-wave function), Tyler
(1973) demonstrated that the liminal Amplitude for reporting beta
apparent motion increases when the Temporal Frequency of oscillation
increases. In his situation, the interstimulus time interval is zero
(ISI = 0); hence an increase in TF means a decrease in the stationary
periods. Korte's fourth law suggests a reciprocity between the
effect of the stationary period and the effect of the ISI. Actually
results of Neuhaus (1930), Sgro (1963), Kahneman (1967), Kahneman
and Wolman (1970) and Kolers (1972) demonstrated that optimal beta
apparent motion is obtained with a constant onset-to-onset time
interval (100 msec < SOA < 120 msec). For stationary periods longer
than some 100 msec., beta motion is obtained with ISI = 0 (Kahneman
and Wolman, 1970). Tyler's data complemented such a conclusion. He
found the lower amplitude limit for beta apparent motion (around
0.2 min of arc) to be constant for stationary periods ranging between 100 to 500 msec (with ISI = 0). For shorter stationary periods,
the amplitude limit for beta motion increased when the stationary
periods decreased. Actually with such an Oscillatory display a
second kind of apparent motion can be observed: the omega motion.
It consists of a "moving shadow" oscillating between two stationary
lights (Saucer, 1953, 1954; Zeeman and Roeloffs, 1953, Tyler, 1973).
The last-named author reported that the proportion of time in which
beta motion occurs decreases regularly with the duration of the stationary periods. For omega motion the proportion first increases
with stationary periods between 500 msec and 100 msec and then decreases when stationary periods decrease as far as 25 msec.

The lower and upper Amplitude limits for the two phenomena
varied differently with the decrease in stationary periods. For the

beta apparent motion, the lower limit increased while the upper limit decreased. For the omega motion, the two limits decreased, but the upper limit decreased faster. Tyler (1973) compared his results with those of Zeeman and Roeloffs (1953) and showed that the upper amplitude limits of beta motion are higher than the upper limits of the omega motion. In other words, the Amplitude-Frequency domain of the beta apparent motion bounds the Amplitude-Frequency domain of the omega apparent motion. Obviously the two phenomena result from a different mode of processing information.

It should be remembered that the Amplitude threshold for Continuous Single motion is bounded by the lower and the upper Amplitude threshold for beta motion with a Discrete Single Motion display (Bonnet, 1975, 1977). In the same line of reasoning Kaufman et al. (1971), using an Oscillatory display, showed that the lower and upper Temporal Frequencies (or velocities) of the Discrete mode for beta motion (8.8 - 21°/sec) bounded the fusion threshold obtained with the Continuous mode (17.3°/sec). In addition, the fusion threshold of the Stop-go-Stop mode is higher (27.9°/sec) than that of the Continuous mode.

Now the conclusion reached is that the shorter the stationary periods, the larger must be the amplitude of the discrete jump or the longer must be the ISI if they are to give raise to an apparent beta motion (see Bonnet, 1975; Kinchla and Allan, 1969). In such circumstances the DAS is assumed to be prominently responsible for the motion sensitivity. Now it could be suggested (mainly because of the results concerning the omega phenomenon) that the MAS also may be at work in the absence of any motion phase. While still largely speculative with regard to the above data, this would suggest that to characterize each system as a set of spatial and temporal filters (cf. King-Smith, 1978a, b) would provide complementary assumptions for explaining these results.

3.3.2. The role of stationary marks

If a stationary mark is introduced, crossing the motion track, different information becomes available: in every condition the organism can use the distance between the two objects as spatial information. Hence such a condition should favour the DAS. Actually, an increase in motion sensitivity due to a stationary mark is considered to be paradigmatically attributable to a prominence of the DAS (Bonnet, 1975, 1977).

With Continuous Single Motion, motion sensitivity (Velocity thresholds) remains unchanged when stationary marks are introduced if exposure time is short enough, or in other words when velocity is high (Leibowitz, 1955b; Hanes, 1965; Mates, 1969; Harvey and Michin, 1974; Bonnet, 1975). Beyond a certain exposure time, however, motion sensitivity increases when a stationary mark is present. The gain in sensitivity increases as exposure time increases. It has been suggested (Bonnet, 1975), although not definitely proved, that the exposure time below which a stationary reference is ineffective and the critical duration (τ) below which the V_{lim} x T trade-off holds are identical.

The proximity of the stationary mark to the motion track is a variable of importance. Using a Discrete mode of presentation with a Single Motion display, Kinchla (1971) showed that motion sensitivity increases with a decrease in the angular separation between the stationary mark and the motion track. This result was obtained with an ISI of either 0.5 sec or 2 sec. Beyond a certain angular separation, stationary marks have no more effect. This critical separation increases with longer exposure times. Harvey and Michon (1974) presented data which complemented those of Kinchla. They used a continuous presentation of a single target which moved toward or away from a similar stationary object. Motion sensitivity, estimated either in terms of liminal Velocity or in terms of liminal Amplitude, decreased when the angular separation between the stationary and the moving targets increased. Their results did not show the asymptotic effect reported by Kinchla (1971). However, the largest separation they used was of 1.6° while the asymptotic effect reported by Kinchla was obtained for a separation of 3°. Nor did Harvey and Michon (1974) observe an effect of a randomly structured background superimposed on their motion configuration. As they themselves suggested, such a stationary background did not add to the information that could be processed.

The effect of stationary marks was also studied by Tyler and Torres (1972) with a Continuous Oscillatory motion display (sine wave function). The liminal Amplitude for detecting such a motion varies curvilinearly with the Temporal frequency of oscillation. Sensitivity is best between 1 and 5 Hz. Without a stationary mark, motion sensitivity increases with a slope of unity (on a log-log plot) between 0.1 and 1 Hz. A stationary mark increases sensitivity. The gain in sensitivity is larger the lower the Temporal Frequency.

3.3.3. Size of the moving target

The effects of changes in the target size of a Single Motion are somewhat puzzling at first sight.

In foveal vision, motion sensitivity, measured in terms of amplitude thresholds, has traditionally been found to decrease with an increase in target area (Bourdon, 1902; Brown, 1931; Graham, 1968; Mates and Graham, 1970). However, Salvatore (1978) showed a better sensitivity for a 2° than for a 0.5° target. Noticeably such a difference was observed in scotopic illumination. It was more reduced in photopic illumination. Barbur and Ruddock (1980) estimating motion sensitivity (through luminance thresholds) to a target, demonstrated a better motion sensitivity for a larger target (8°) than for a smaller one (1.7°). They also found that the higher the velocity of the translation, the larger the effect of target size. Actually, it was found (Barbur, personal communication) that for small targets (0.6°) motion sensitivity tends to decrease with velocity while for larger ones it tends to increase towards an asymptotic value.

In peripheral vision, McColgin (1959, 1960), Salvatore (1978)

and Barbur and Ruddock (1980) showed an increase in motion sensitivity with an increase in target size.

Close examination of Barbur and Ruddock's data (1980) suggests
an explanation for some of the reported results. In the very low
velocity range, there may be some indication of better motion sensitivity with smaller target sizes.

Actually the effect of target size may be critical for determining in a given experimental situation which process is more prominent: the MAS or the DAS. It was suggested (Bonnet, 1975), that
an increase in motion sensitivity with an increase in target size
should indicate a spatial summation process. Alternatively, a decrease in motion sensitivity with an increase in target size would
result from a greater uncertainty in target position, which would
consequently raise the liminal amplitude for motion detection if
this dimension is going to be processed.

Results obtained with travelling sine-waves enlighten the
relationship between the size of a target and motion sensitivity.
With low velocity drifts (low TF) motion sensitivity varies curvilinearly with a decrease in the period of the grating. Hence a
low Spatial Frequency grating can raise a lower motion sensitivity
than can a higher Spatial Frequency grating. In contrast with fast
velocity drifts, motion sensitivity is a low pass function of
Spatial Frequency, in such a way that motion sensitivity is always
better for a large period grating than for a small period one (i.e.
high SF).

3.3.4. Velocity thresholds of drifting gratings

The V_{lim} x T trade-off has always been demonstrated with targets
having a high luminance contrast with regard to their background.
Several failures of the V_{lim} x T trade-off have been reported. Using
a Single motion target, Henderson (1971, 1973) and Bonnet (1975)
observed first a decline in A_{lim} (= V x T) with increased exposure
time (T). Beyong a certain exposure time A_{lim} increases with T.
Such failures of the Velocity x Time trade-off have been attributed
to some luminance bias in the display (Bonnet, 1975): for short
exposure times the luminance of the oscilloscope spot, which served
as the target, did not reach its nominal value within the exposure
time. Consequently the real luminance of the spot increased with
exposure time. The data reported in Figure 4 confirm such an interpretation. In the present replication (see § 3.2., Fig. 4) of the
1975 experiment, neutral density filters were used to keep the mean
luminance of the spot constant at every exposure time. These results
now show the Velocity x Time trade-off when T ≤ 180 msec.

Recently van der Glas et al. (1979) presented data which fail
to demonstrate such a V_{lim} x T trade-off in a condition where a
luminance bias is excluded. These authors presented low-contrast
square-wave drifting gratings flashed on for a duration ranging
between 25 to 6400 msec. The liminal Velocity for motion detection
was measured. Van der Glas et al. reported that when V_{lim} x T is
plotted on log-log coordinates as a function of the exposure time

(T), the function showed a U shape. The descending part of the function represented a failure of the V_{lim} x T and is an unexpected result. If the V_{lim} x T trade-off is taken for granted, a possible explanation for this failure may be the use of very low contrast gratings. In effect, a trade-off between exposure time (T) and Modulation contrast has been demonstrated for stationary gratings (see § 3.1.1.). Preliminary results of the present author also suggest such a trade-off for motion detection. Actually, van der Glas et al. showed that V_{lim} decreases with an increase in contrast until an asymptotic value. Hence, in their conditions, two summations could have been combined in such a way that looking for only one may lead to a conclusion of its failure. This speculation should be tested in further experiments.

4. EFFECTS OF RETINAL ECCENTRICITY

Several times it has been stated that the peripheral retina is specialized for motion detection (Exner, 1886; Granit, 1930; Sharpe, 1974). In general, visual sensitivity is found to decline with eccentricity (van Doorn et al., 1972; Hilz and Cavonius, 1974; van der Wildt et al., 1976; Kroon et al., 1980; Rijsdijk et al., 1980). The assumed specialisation of the peripheral retina for motion detection could be understood in different ways. Either motion sensitivity is better in the peripheral retina than in the fovea or motion sensitivity is better than sensitivity for a stationary pattern in the periphery, while the reverse would be true in the fovea. A third interpretation would suggest that when the MAS is prominent, motion sensitivity is independent from eccentricity.

Many of the results using a Single Motion or repetitively moving target, demonstrated that velocity or amplitude thresholds for motion detection increase with eccentricity in much the same way as visual resolution does for stationary targets (Aubert, 1887; Basler, 1906; Klein, 1942; Warden et al., 1945; Gordon, 1947; McColgin, 1959, 1960; Leibowitz et al., 1972; Johnson and Leibowitz, 1976; Salvatore, 1978).

Many factors could explain the poorer performance of the peripheral retina. Among them, the refractive error of the eye has been invoked by Granit (1930). Leibowitz, Johnson and Isabelle (1972) show convincingly that correction of the refractive error at every eccentricity, leads to more homogeneous velocity thresholds between subjects. It also improves motion sensitivity with increasing eccentricity. Without correction, over a range of 80° of eccentricity, motion thresholds increased by an average factor of 10. After correction for refractive error, they only increased by a factor of 6. In this situation, the mode of presentation (stop-go-stop), the exposure time (1 sec) and the size of the target (1°) suggested the DAS to be prominent. The importance of a dioptric factor may be related to such prominence. Arguments will be presented below suggesting that dioptric factors do not affect MAS sensitivity.

4.1. Psychophysical receptive fields and the Magnification factor

One other explanation of the decrease in sensitivity with retinal eccentricity lies in the fact that the mean neural receptive
field size increases with eccentricity (Hubel and Wiesel, 1960,
1974). Different paradigmatic techniques were used to estimate psychophysically the receptive field sizes at different eccentricities:
the point or line spread function estimated from the Fourier transform of the Modulation Transfer Function (Hilz and Cavanius, 1974),
the direct use of this MTF (Koenderink et al., 1978a, b; Rovamo
et al., 1978; Kroon et al., 1980; Rijsdijk et al., 1980) and
Westheimer's paradigm (Ransom-Hogg and Spillman, 1980). In every
case the estimated receptive field size increases monotonically
with eccentricity. Quantitavely, there is a large spread in the
estimates (see Ransom-Hogg and Spillman, 1980). Now questions have
been asked as to whether there is a unique size of the receptive
fields at every eccentricity (van Doorn et al., 1972; van der Wildt
et al., 1976) or whether there are several receptive field sizes
at every eccentricity (Graham et al., 1977; Koenderink, 1977). A
reasonable assumption is proposed by Koenderink (1977) who speculat-
es that there may exist a large distribution of receptive field
sizes at every eccentricity, but with a shift in the lower limit of
the receptive field sizes.

Hence a shift in the mean receptive field size would occur,
which would explain the shift in resolution of the visual system
as a function of eccentricity.

Apart from the increase in the mean size of receptive fields,
other factors may contribute in a related way to the decline in
sensitivity with eccentricity: the decrease in the density of receptors and the increase in synaptic convergence (Granit, 1930) or
the decrease in ganglion cell density (Wilson and Sherman, 1976).
When all these factors are taken into consideration they lead to
the definition of a cortical magnification factor, which would represent the extent of the relationship between one degree on the
retina and the linear distance of the cortex where it is mapped.
Such a Magnification factor (M in mm/deg.) could be related to the
square root of ganglion receptive field density (Wilson and Sherman,
1976; Rovamo and Virsu, 1979). In other words the inverse Magnification factor (M^{-1} in deg./mm) would increase linearly with mean receptive field sizes.

There are some uncertainties in estimating this Magnification
factor (see Rovamo and Virsu, 1979; Ransom-Hogg and Spillman, 1980).
Nevertheless, when it is taken into consideration, the modulation
thresholds for stationary or for drifting gratings can be expressed
as a function of a Cortical Spatial Frequency (in cycles mm^{-1}).
Estimates showed that in photopic vision such a transformation leads
to a unique Modulation Transfer Function of the visual system irrespective of eccentricity (Rovamo et al., 1978; Virsu and Rovamo,
1979; Koenderink et al., 1978a, b). Some deviations from the scaled
functions have appeared for the high Spatial Frequencies (Virsu and

Rovamo, 1979). However, they may be accounted for by optical factors
(Campbell and Green, 1965).

The cortical Magnification factor assumption would be validated
in "suitable temporal conditions" (Rovamo et al., 1978). These temporal conditions appeared to be either a 500 msec exposure time for
stationary gratings or a constant and slow velocity (0.78°/sec) for
drifting gratings (Virsu and Rovamo, 1979, their Fig. 2). However,
the assumption holds essentially for motion detection of a grating
at 4 Hz (Rovamo et al., 1978; Virsu and Rovamo, 1979, their Fig. 3).
In a first approximation, it could be suggested that the Magnification factor assumption holds for motion detection in conditions in
which the DAS is prominent. Unfortunately for the latter assumption,
the authors have not systematically explored the Temporal Frequency
(or Velocity) domain such as to define the limit of validity of
their assumption. Koenderink et al. (1978b) stated in effect that
the best sensitivity for moving gratings is obtained with a *unique*
combination of Spatial Frequency and Temporal Frequency. Tyler and
Torres (1972) demonstrated an asymmetry in the effect of eccentricity
as a function of the range of the Temporal Frequencies used with
Oscillatory motion: for Temporal Frequencies ranging between 0.2 to
5 Hz, the reciprocal liminal Amplitude ($1/A_{lim}$) increases with TF
with a slope of unity in the fovea and at 20° eccentricity. For that
range of TF's, eccentricity seems only to produce a shift in the
intercept of the sensitivity function. For the upper TF range the
effect of eccentricity is more accentuated: sensitivity ($1/A_{lim}$)
decreases with an increase in Temporal Frequency. The decline is
steeper in peripheral vision that in the fovea and at each TF the
difference in sensitivity for different eccentricities is larger
than it is in the low TF range. Keeping this theoretical issue in
mind, let us summarize the main effects of eccentricity on motion
sensitivity to drifting gratings. Koenderink et al. (1978a, b) have
run an extensive series of experiments which reported most of the
known results about this question:

    - For a constant size of the moving field, the upper cut-off
of the sensitivity function declines monotonically with the increase
in eccentricity even in the low range (0° to 8°).

    - With a constant drift rate (2 Hz), an increase in eccentrici-
ty reduces the overall sensitivity, shifts the optimum toward lower
Spatial Frequencies and reduces the low SF attenuation.

    - At any eccentricity, the optimum sensitivity is given by a
unique combination of Spatial and Temporal Frequencies: when eccentricity increases from 0° to 50° the optimal SF decreases, while the
Temporal Frequency increases. Correspondingly the optimal velocity
increases with eccentricity, from about 2°/sec in the fovea to
12°/sec with 50° eccentricity.

    - Enlarging the moving field size from 0.5° to 16° reduces the
deterioration of sensitivity with eccentricity. With a 0.5° square
field the modulation threshold doubles every 2.3° of eccentricity.
With a 4° field this threshold doubles every 12°. If the field is
large enough the modulation thresholds are equivalent in periphery

and fovea, but they are shifted to lower Spatial Frequencies.

- In foveal vision the amplitude modulation thresholds (AL)
increase as the square-root of the average luminance (L) according
to the de Vries-Rose law for Spatial Frequency. They range between
0.25 and 1.5 cpd (with a TF = 4 Hz). At 50° eccentricity for the
same SF's and TF, the amplitude modulation thresholds (AL) increase
linearly with the average luminance (L) according to the Weber law.

4.2. Effects of stationary marks and stationary positions

Improvement of sensitivity due to the presence of stationary
marks close to the motion track and/or to the presence of stationary
positions in the target translation path has been assumed to paradigmatically prove the prominence of the DAS. Consequently, since the
MAS is believed to be prominent in peripheral vision, the presence
of such "stationary information" should be ineffective if the two
systems are independent.

In foveal vision, Bonnet and Renard (1977) have shown that the
Amplitude thresholds are generally lower for Discrete and Stop-go-
Stop modes of presentation than for Continuous Single Motion. At 30°
eccentricity along the vertical meridian lower Amplitude thresholds
are now obtained with Continuous Single Motion. Higher thresholds
are obtained with the Stop-go-Stop mode of presentation, suggesting
that the stationary positions may interfere with the processing of
the motion phase by the MAS. Much higher thresholds are obtained
with the Discrete mode.

Tyler and Torres (1972) showed that although in foveal vision
a stationary mark decreases the Amplitude thresholds of an Oscillatory motion with a frequency below about 5 Hz, the thresholds remain
unchanged in peripheral vision.

4.3. The peripheral retina specialized for motion detection

Next, results will be reported which show that on some occasions
motion sensitivity can be independent of eccentricity. Bonnet and
Renard (1977) found that the Amplitude (or velocity) threshold for
a Continuous Single motion of a 5 min light spot was the same in
central vision and at 30° of eccentricity along the vertical meridian (T = 360 msec).

The luminance adaptation level was scotopic. In such conditions
the A_{lim} of a Stop-go-Stop Single motion is 1.3 times larger in the
periphery than in central vision. Finally the A_{lim} for a Discrete
Single motion is 2.8 times larger. At that exposure time the amplitude threshold is similar for the three modalities of this Single
motion in central vision (about 10 min of arc).

At a shorter exposure time, A_{lim} is shorter for Discrete Single
motion. This result may suggest that in peripheral vision a stationary position of a target leads to some inhibition of motion de-

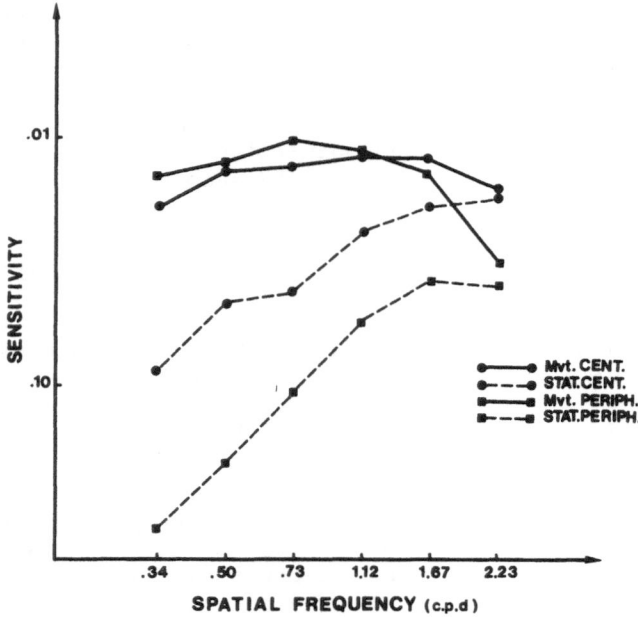

Fig. 5. Sensitivity for motion detection of drifting sine-wave
gratings and for stationary gratings in central vision and
at 60° of eccentricity in the nasal meridian (Bonnet and
Chaudagne, 1979a).

tection for which the MAS is predominantly responsible. Further experiments should explore this point systematically.

Barbur and Ruddock (1980) also found that at 30° eccentricity
in the temporal retina the contrast (luminance) threshold may be
lower than in foveal vision for a target moving at 17.5°/sec through
a 8° track.

Using Frequency motion, Bonnet and Chaudagne (1979b) attempted
to show in which conditions motion sensitivity is invariant with
respect to eccentricity. They chose low spatial frequency gratings
(between .34 to 2.23 cpd) with a low mean luminance (.1 cd.m^{-2}) and
a scotopic adaptation level (.01 cd.m^{-2}) presented in a large
square field (6.7°) with a nearly optimal Temporal Frequency drift
(5 Hz). Modulation thresholds for the detection of motion of the
grating were measured. In such conditions as illustrated in Fig. 5,
sensitivity for motion detection is nearly equal in central vision
and at 60° eccentricity along the nasal meridian. However, contrast
sensitivity for stationary gratings always remains lower in peripheral than in central vision. Actually these conditions were chosen
in order to confirm inferences derived from previous experiments.
In effect four main factors are believed to contribute to the con-

stant motion sensitivity with eccentricity:

a) Temporal Frequency of drift should be optimal in order to reduce the low Spatial Frequency attenuation at best (Sharpe and Tolhurst, 1973; Koenderink et al., 1978a). Such a factor is consistent with the high velocities at which the results were obtained by Bonnet and Renard (1977) and by Barbur and Ruddock (1980).

b) Low Spatial Frequencies for which sensitivity is less affected by eccentricity (Koenderink et al., 1978b). It should also be stressed that dioptic factors and accommodation affect to a much lesser extent sensitivity to low SF (Campbell and Green, 1965). Remember also the large size of the target in Barbur's experiment.

c) A large motion field. Koenderink et al. (1978b, their Figs. 1 and 3) demonstrated that an increase in the target field size improves the sensitivity in peripheral vision more than in the central retina.

d) Low retinal luminance. Koenderink et al. (1978c, p. 861), and Virsu and Rovamo (1979) showed that a decrease in retinal luminance does mainly reduce the sensitivity of the central retina but scarcely affects the sensitivity of the peripheral retina, at least as far as low SF are concerned (Daitch and Green, 1969). Such a finding is termed a central scotopic phenomenon in the literature on perimetry.

In conclusion, even though the increase in eccentricity selectively changes the relative prominence of the MAS and the DAS, it appears that it never excludes the DAS. Motion sensitivity never appears to be significantly better in the peripheral than in the central retina. In some extreme cases, motion sensitivity can be equal at different eccentricities. This is probably the way in which the specialisation of the peripheral retina for motion sensitivity should be understood.

4.4. Self motion sensitivity

There is a motion experience which separates the central and peripheral retinal functions to a larger extent. This is the case with self-motion sensations. Self-motion experience, also called vection, is generated in a stationary observer while some pattern is moving in his peripheral field of vision (see the chapters by Berthoz and Droulez and by Leibowitz et al. in this volume; Dichgans and Brandt, 1978). Such an experience is assumed to result from a two step processing, first in visual areas, then in central structures that normally process vestibular inputs (Dichgans and Brandt, 1978, p. 781).

Measuring motion thresholds for self-motion experience is a first attempt to answer the question which of the two visual system (MAS or DAS) pre-processes the visual inputs that are further elaborated in vestibular structures.

Berthoz et al. (1975) measured the velocity thresholds of a pattern drifting in the peripheral visual field. They used two

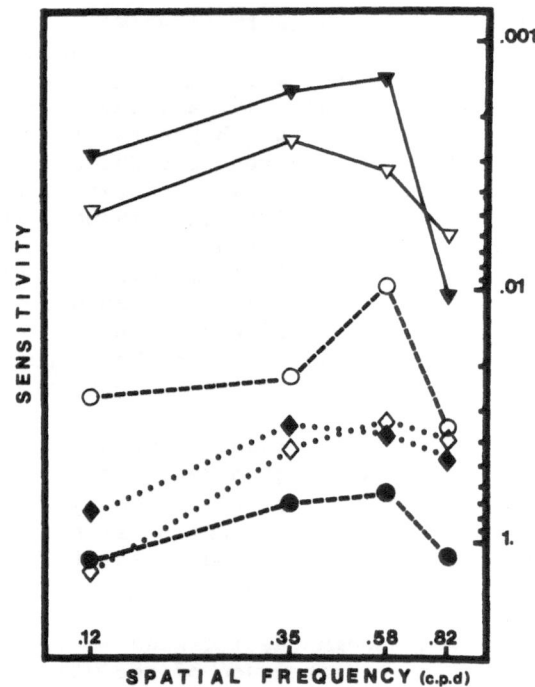

Fig. 6. Sensitivity for Object-motion ($\triangledown$, $\blacktriangledown$); Self-motion ($\diamondsuit$,$\blacklozenge$),
and Stationary sine-wave gratings (O , ●) at 60° eccentri-
city; open and dark symbols stand for each of two subjects
(Bonnet and Chaudagne, 1979b).

motion sensation criteria: either object-motion or self-motion sensation with unlimited exposure time. Velocity thresholds are rather
low and equal for the two sensation criteria. Because of their low
level (slow velocity), a prominence of the DAS is suggested by these
data.

Berthoz et al. (1975) and Leibowitz et al. (1979) measured luminance thresholds for vection. The former used sagittal (linear)
self-motion, while the latter used circular self-motion. They both
concluded that the luminance thresholds are very close to those for
image detection, but they did not attempt to measure object-motion
thresholds in similar conditions. Bonnet and Chaudagne (1979a) took
this problem up. They used sine-wave drifting gratings presented at
60° eccentricity in a 22° x 22° field. Four Spatial Frequencies were
used ranging between .12 and .82 cpd. Three different modulation
thresholds were measured in separate conditions:

a) thresholds for the detection of the orientation (vertical
vs. horizontal) of stationary gratings. These thresholds estimate

pattern sensitivity.

b) thresholds for the detection of the direction of object-motion (either leftward vs. rightward or upward vs. downward).

c) thresholds for the detection of the self-motion sensation, the direction of which is opposite to that of the object motion.

The temporal frequency of the drifting gratings was 4 Hz. As shown in Fig. 6, for the two experienced subjects taking part in the experiment no significant difference between self-motion sensitivity and pattern sensitivity could be demonstrated while sensitivity for object-motion is 1 log unit better.

Hence it would appear from these results that pattern visibility is a prerequisite for self-motion experience. However, preliminary observations with subjects having very degraded pattern vision indicate that they may have nearly normal self-motion sensitivity. If these results are confirmed, they would suggest that the MAS is actually preprocessing visual inputs before self-motion sensation is elaborated in central vestibular structures.

5. CONCLUSIONS

The present review has presented rather integrative points of view upon motion thresholds. First, instructions should define the task in terms of the detection of a directional component in the stimulus, irrespective of other phenomenal appearances. This argument holds only in case two opposite directions of motion are used (see also Sekuler et al., in this volume). Second, motion sensitivity can equivalently be estimated on the basis of one of the several dimensions which define a stimulus. Liminal Velocity, Amplitude, Exposure-Time, Temporal-Frequency, Luminance Contrast and Modulation are the most frequently used. Trade-off functions have been demonstrated between several pairs of these dimensions, such as Velocity x Time, Spatial Frequency x Velocity.

Some other trade-off functions should be more thoroughly worked out, as for instance Size x Luminance (see Bonnet, 1975), and several trade-off functions taking either contrast or modulation into account. It is likely that a trade-off between two dimensions can only be demonstrated when other variables are beyond the limiting value for which they also show a trade-off.

A conceptual background has been proposed in terms of two Analysing Systems processing different aspects of the spatio-temporal information (Bonnet, 1977). It synthetizes the common aspects of different models present in the literature by assuming their basic equivalence. Rules of functioning can be described for each of these two systems, consistent with the type of motion display used in the experiment. Although these two Systems have to be conceived as separate, they need not be assumed to be independent. Their separation is shown in the rules of processing as can be inferred from experimental results.

Two analysing systems having large overlapping domains (see for

instance Kulikowski, 1978; King-Smith, 1978a, b) are assumed to be
at work. They are functionally separate but not independent. In
most of the conditions it is not easily decidable which of these two
Systems is responsible for the overt response of movement detection.
It is only in some extreme cases that the response may be attributed
to a single System. However, in successively approximating the
psychophysical functions which should be characteristic of a given
System, it becomes possible to decide on their relative prominence
in most of the conditions. For motion detection, these two systems
have to process spatio-temporal modulation with a directional component.

The Movingness Analysing System is likely to be as effective
in the periphery as in central vision. The MAS would process mainly
those intensive changes (i.e. velocity) which have a directional
component. Such a statement is consistent with Exner's assumption
about the role of the velocity range. Whether the results are expressed in the Space-Time domain or in the Spatial Frequency-
Temporal Frequency domain, the MAS can be described as made of
velocity-tuned units (see Orban, 1977) or as made of a low-pass
Spatial Frequency filter and of a band-pass Temporal Frequency
filter, depending upon which functional aspect is going to be
stressed. In every case, the MAS can psychophysically be characterized by the fact that Spatial and Temporal summation laws hold.
It seems clear that in contradiction with some statements (for
instance Henderson, 1971, 1973), these laws do not necessitate requirements for target visibility even if they appear to be parallel
to visibility functions (see Bouman and van den Brink, 1953; van
den Brink and Bouman, 1957). In effect, the temporal limits for the
V_{lim} x T trade-off in a Single motion display are much higher than
those required for target visibility. Such an argument will be re-
inforced if it is demonstrated that the failure in van der Glas
et al. (1978) to show such a trade-off is due to the use of a very
low contrast grating.

The Displacement Analysing System is clearly more efficient in
central vision. The DAS would process mainly extensive changes
having a directional component. Hence, it would mostly come into
play with low velocity motion displays. The DAS could either be described as made of amplitude-tuned units (see Orban, 1977) or as
made of a band-pass Spatial Frequency filter and of a low-pass Temporal Frequency filter. The question of whether such filters are
broad-band unique filters or the envelope of selectively tuned
channels is beyond the scope of the present paper. Psychophysically,
the DAS has been characterized by the efficiency with which stationary components lower motion thresholds. Such a study has not
been emphasized with Frequency motion displays, which almost always
contain stationary marks such as the fixation point or the frame
in which the gratings are presented.

The assumption of two separate Analysing Systems for motion
detection has been related to neurophysiological findings. For most
of the authors (see Kulikowski and Tolhurst, 1973; King-Smith,

1978a, b) these two psychophysically characterized systems are consistent with the two way classification of ganglion cells in higher
mammals (Enroth-Cugell and Robson, 1966; Ikeda and Wright, 1975; de
Monasterio and Gouras, 1975). However, there are some inconsistencies
between the psychophysical criteria and the neurophysiological criteria of the transient and sustained systems (MacLeod, 1978; Sekuler
et al., 1978; Wilson, 1980; Lennie, 1980). Moreover, in most of the
psychophysical studies devoted to that problem flicker and movement
have been confused. However, Movshon et al. (1978) provided convincing neurophysiological data for such an interpretation. Another
line of speculation about the physiological substratum of these two
systems is provided by the classification of the two visual systems
(Schneider, 1968; Trevarthen, 1968; Bonnet, 1975). The study of
subjects lacking afference to the visual cortex and who have nearly
normal motion thresholds supports such speculations (Barbur et al.,
1980). Now a third line of arguments is provided by the results of
Orban (1975, 1977 and 1981) who demonstrated in area 18 of the cat
that while some directional cells are velocity-selective, some
others are amplitude-selective. The fact that three different but
related kinds of neurophysiological substrata can be proposed for
two psychophysically defined systems stresses that such an approach
is necessarily more integrative. A correspondence between these
systems should not be expected to be strict although, from a
heuristic point of view, it is a fertile point of departure.

ACKNOWLEDGEMENT

    The author wishes to thank Dr. L. Ganz, Dr. R. Sekuler,
Dr. H.W. Leibowitz and Dr. A.H. Wertheim for their valuable
comments and their help in the improvement of the English.

REFERENCES

Aarons, L., Visual apparent movement research: review 1935-1955
        and bibliography 1955-1963. Perceptual and Motor Skills,
        1964, 18, 230-274.
Anstis, S.M., Apparent movement. In: R. Held, H.W. Leibowitz and
        H.-L. Teuber (Eds.). Handbook of Sensory Physiology, Vol. VIII,
        Berlin, Springer-Verlag, 1973.
Arend, L.E., Response of the human eye to spatially sinusoidal
        gratings at various exposure durations. Vision Research, 1976,
        26, 1311-1315.
Arend, L.E. and Lange, R.W., Influence of exposure duration on the
        tuning of spatial channels. Vision Research, 1979, 19, 195-201.
Aubert, B., Die Bewegungsempfindung. II Mitt. Archiv für die
        gesamte Physiologie, 1887, 40, 450-473.
Barbur, J.L. and Ruddock, H.H., Spatial characteristics of movement
        detection mechanisms in human vision. I. Achromatic vision.

Biological Cybernetics, 1980, 37, 77-92.

Barbur, J.L., Holliday, I.E., Ruddock, K.H. and Waterfield, V.A.,
    Spatial characteristics of movement detection mechanisms in
    human vision. III. Subjects with abnormal visual pathways.
    Biological Cybernetics, 1980, 37, 99-105.

Basler, A., Über das Sehen von Bewegung. I. Mitt. Die Wahrnehmung
    kleinster Bewegungen. Archiv für die gesamte Physiologie, 1906,
    115, 582-601.

Berthoz, A., Pavard, D. and Young, L.R., Perception of linear
    horizontal self-motion induced by peripheral vision (linear
    vection). Experimental Brain Research, 1975, 23, 471-489.

Bonnet, C., Les mécanismes de la perception d'un mouvement visuel.
    Bulletin de Psychologie, 1971, 24, 415-417.

Bonnet, C., A tentative model for visual motion detection.
    Psychologia, 1975, 18, 35-50.

Bonnet, C., Visual motion detection models: features and frequency
    filters. Perception, 1977, 6, 491-500.

Bonnet, C. and Chaudagne, N., Comparison of contrast thresholds for
    object-motion and for self-motion in different directions.
    Experimental Brain Research, 1979, 36, R6(a).

Bonnet, C. and Chaudagne, N., Relative sensitivity for moving vs.
    stationary gratings in central and peripheral vision. Paper
    read at the 2nd European Conference on Visual Perception,
    Noordwijkerhout (The Netherlands), 1979(b).

Bonnet, C. and Renard, C., La détection du mouvement visuel en
    vision centrale et en vision périphérique. In: Psychologie
    expérimentale et Comparée; hommage à Paul Fraisse, Paris,
    P.U.F., 1977.

Bouman, M.A. and Brink G., van den, Absolute thresholds for moving
    point sources. Journal of the Optical Society of America,
    1953, 43, 895-898.

Bourdon, B., La perception visuelle de l'espace. Paris: Reinwald,
    1902.

Breitmeyer, R.G. and Ganz, L., Temporal studies with flashed
    gratings: inference about human transient and sustained
    channels. Vision Research, 1977, 17, 861-866.

Brink, G., van den and Bouman, M.A., Visual contrast threshold for
    moving point sources. Journal of the Optical Society of
    America, 1957, 47, 612-618.

Brown, J.F., The thresholds for visual movement. Psychologische
    Forschung, 1931, 14, 249-268.

Brown, R.H., Velocity discrimination and the intensity time relation.
    Journal of the Optical Society of America, 1955, 45, 189-192.

Brown, R.H., Influence of stimulus luminance upon the upper speed
    threshold for the visual discrimination of movement.
    Journal of the Optical Society of America, 1958, 48, 125-128.

Caelli, T., Hoffman, W.C. and Lindman, E., Apparent motion: self-
    excited oscillations induced by retarded neural flows.
    In: Leeuwenberg and Buffart (Eds.). Formal theories of visual
    perception, New York, Wiley, 1978.

Campbell, F.W. and Green, D.G., Optical and retinal factors affecting visual resolution. Journal of Physiology, 1965, 181, 576-593.

Campbell, F.W. and Robson, J.G., Application of Fourier analysis to the visibility of gratings. Journal of Physiology, 1968, 197, 551-556.

Cohen, R.L. and Bonnet, C., Movement detection thresholds and stimulus durations. Perception and Psychophysics, 1972, 12, 269-272.

Crook, M.N., Visual discrimination of movement. Journal of Psychology, 1937, 3, 531-558.

Daitch, J.M. and Green, D.G., Contrast sensitivity of the human peripheral retina. Vision Research, 1969, 9, 947-952.

Dichgans, J. and Brandt, Th., The visual-vestibular interaction: Effects on self-motion perception and postural control. In: R. Held, H.W. Leibowitz and H.L. Teuber (Eds.). Handbook of sensory Physiology, Vol. VIII, Berlin. Springer-Verlag, 1978.

Dimmick, F.L. and Karl, J.C., The effect of exposure time upon the R.L. of visual motion. Journal of Experimental Psychology, 1930, 13, 365-369.

Doorn, A.J. van, Koenderink, J.J. and Bouman, M.A., The influence of retinal inhomogeneity on the perception of spatial patterns. Kybernetik, 1972, 10, 223-230.

Enroth-Cugell, C. and Robson, J.G., The contrast sensitivity of retinal ganglion cells of the cat. Journal of Physiology, 1966, 187, 517-522.

Exner, S., Über das Sehen von Bewegungen und die theorie des zusammengesetzten Auges. Sitzungsberichts Akademie Wissenschaft Wien, 1875, 72, 156-190.

Exner, S., Ein Versuch uber die Netzhaut Perripherie als Organ zur Wahrnehmung von Bewegungen. Archiv für die Gesamte Psychologie, 1886, 38, 217-218.

Exner, S., Über optische Bewegungsempfindungen. Biologisches Zetralblatt, 1888, 8, 14.

Ganz, L., Temporal factors in visual perception. In: E.C. Garterette and M.F. Friedman (eds.). Handbook of Perception, Vol. V, Seeing, New York. Academic Press, 1975.

Glas, H.W. van der, Orban, G.A., Joris, Ph.X. and Verhoeven F.J., Velocity thresholds measured with low contrast gratings. Paper presented at the 2nd European Conference on Visual Perception. Noordwijkerhout (The Netherlands), 1979.

Gordon, D.A., The relation between the thresholds of form motion and displacement in parafoveal and periferal vision at a scotopic level of illumination. American Journal of Psychology, 1947, 60, 202-225.

Gorea, A., Le traitement visual des fréquences spatiales. These de 3eme cycle. Université René Descartes. Paris, 1978.

Gorea, A., Directional and nondirectional coding of a spatio-temporal modulated stimulus. Vision Research, 1979, 19, 545-551.

Graham, C.H. (Ed.), Vision and visual perception. New York, John
    Wiley and Sons, 1965.
Graham, C.H., Depth and movement. American Psychologist, 1968, 23,
    18-26.
Graham, N., Robson, J.G. and Nachmias, J., Grating summation in
    fovea and periphery. ARVO Spring Meeting, Sarasota, 1977.
Granit, R., Comparative studies on the peripheral and central retina
    I. An interaction between distant areas in the human eye.
    American Journal of Physiology, 1930, 94, 41-50.
Hanes, L.F., Discrimination of direction of movement at short ex-
    posure durations. Dissertation Abstracts International, 1965,
    25, 216-219.
Harvey, L.O. and Michon, J.A., The detectability of relative motion
    as a function of exposure duration angular separation and back-
    ground. Journal of Experimental Psychology, 1974, 103, 317-325.
Henderson, D.C., The relationship among time distance and intensity
    as determinants of motion discrimination. Perception and Psycho-
    physics, 1971, 10, 313-320.
Henderson, D.C., Visual discrimination of motion: stimulus relation-
    ship at threshold and the question of luminance-time recipro-
    city. Perception and Psychophysics, 1973, 13, 121-130.
Hilz, R. and Cavonius, C.R., Functional organization of the peri-
    pheral retina: sensitivity to periodic stimuli. Vision
    Research, 1974, 14, 1333-1337.
Hubel, D.H. and Wiesel, T.N., Receptive fields of the optic nerve
    fibers in the spider monkey. Journal of Physiology, 1960,
    154, 572-580.
Hubel, D.H. and Wiesel, T.N., Uniformity of monkey striate cortex:
    A parallel relationship between field size scatter and magnifi-
    cation factor. Journal of Comparative Neurology, 1974, 158,
    295-306.
Ikeda, H. and Wright, M.J., Spatial and temporal properties of
    "sustained" and "transient" neurons in area 17 of the cat's
    visual cortex. Experimental Brain Research, 1975, 22, 363-383.
Johnson, C.A. and Leibowitz, H.W., Velocity-time reciprocity in the
    perception of motion: foveal and peripheral determinations.
    Vision Research, 1976, 16, 177-180.
Kahneman, D., An onset-onset law for one case of apparent motion
    and metacontrast. Perception and Psychophysics, 1967, 2,
    577-584.
Kahneman, D. and Wolman, R.E., Stroboscopic motion: effects of
    duration and interval. Perception and Psychophysics, 1970, 8,
    161-164.
Kaufman, L., Cyrolnik, I., Klapowitz, J., Melnick, G. and Stoff, D.,
    The complementarity of apparent and real motion. Psychologische
    Forschung, 1971, 34, 343-348.
Keesey, U.T., Flicker and pattern detection: a comparison of thres-
    holds. Journal of the Optical Society of America, 1972, 62,
    440-448.

Kelly, D.H., Adaptation effects on spatio-temporal sine-wave thresholds. Vision Research, 1972, 12, 89-101.

Kelly, D.H., Visual contrast sensitivity. Optica Acta, 1977, 24, 107-129.

Kelly, D.H., Motion and vision. II. Stabilized spatio-temporal threshold surface. Journal of the Optical Society of America, 1979, 69, 1340-1349.

Kinchla, R.A., Visual movement perception: a comparison of absolute and relative movement discrimination. Perception and Psychophysics, 1971, 9, 165-171.

Kinchla, R.A. and Allan, L.G., A theory of visual movement perception. Psychological Review, 1969, 76, 537-558.

King-Smith, P.E., Visual sensitivity to moving stimuli: data and theory. In: J.C. Armington, J. Krauskopf and B.R. Wooten (Eds.) Visual psychophysics and physiology, New York. Academic Press, 1978a.

King-Smith, P.E., Analysis of the detection of a moving line. Perception, 1978b, 7, 449-458.

King-Smith, P.E. and Riggs, L.A., Visual sensitivity to controlled motion of a line or edge. Vision Research, 1978, 18, 1509-1520.

King-Smith, P.E., Riggs, L.A., Moore, R.K. and Butler, T.W., Temporal properties of the human visual nervous system. Vision Research, 1977, 17, 1101-1106.

Klein, G.S., The relation between motion and form acuity in parafoveal and peripheral vision and related phenomena. Archivs of Psychology, 1942, 39, 1-70.

Koenderink, J.J., Current models of contrast processing. In: R. Spekreyse and L.H. van der Tweel (Eds.). Spatial Contrast, Amsterdam, North-Holland, 1977.

Koenderink, J.J., Bouman, M.A., Bueno de Mesquita, A.E. and Slappendel, S., Perimetry of contrast detection thresholds of moving spatial sine wave patterns. II. The far peripheral visual field. Journal of the Optical Society of America, 1978a, 68, 850-854.

Koenderink, J.J., Bouman, M.A., Bueno de Mesquita, A.E. and Slappendel, S., Perimetry of contrast detection thresholds of moving spatial sine wave patterns. III. The target extent as a sensitivity controlling parameter. Journal of the Optical Society of America, 1978b, 68, 854-859.

Koenderink, J.J., Bouman, M.A., Bueno de Mesquita, A.E. and Slappendel, S., Perimetry of contrast detection thresholds of moving spatial sine wave patterns. IV. The influence of the mean retinal illuminance. Journal of the Optical Society of America, 1978c, 68, 860-865.

Koenderink, J.J. and Doorn, A.J. van, Spatio temporal contrast detection threshold surface is bimodal. Optics Letters, 1979, 4, 32-34.

Kolers, P.A., Aspects of motion perception. International series of monographs in experimental psychology. Vol. 16, Oxford, Pergamon Press, 1972.

Korte, A., Kinematoskopische untersuchungen. Zeitschrift für Psychologie, 1915, 72, 193-296.

Krauskopf, J., Effect of retinal image motion on contrast thresholds for maintained vision. Journal of the Optical Society of America, 1957, 47, 740-744.

Kroon, J.N., Rijsdijk, J.P. and Wildt, G.J. van der, Peripheral contrast sensitivity for sine-wave gratings and single period. Vision Research, 1980, 20, 243-252.

Kulikowski, J.J., Effect of eye movements on the contrast sensitivity of spatio-temporal patterns. Vision Research, 1971, 11, 261-274.

Kulikowski, J.J., Spatial resolution for the detection of pattern and movement (real and apparent). Vision Research, 1978, 18, 237-238.

Kulikowski, J.J. and Tolhurst, D.J., Psychophysical evidence for sustained and transient channels in human vision. Journal of Physiology, 1973, 232, 149-163.

Legge, G.E., Sustained and transient mechanisms in human vision: temporal and spatial properties. Vision Research, 1978, 18, 69-81.

Leibowitz, H.W., The relationship between the rate threshold for the perception of movement and luminance for various durations of exposure. Journal of Experimental Psychology, 1955a, 49, 209-214.

Leibowitz, H.W., Effect of reference lines on the discrimination of movement. Journal of the Optical Society of America, 1955b, 45, 829-830.

Leibowitz, H.W., Johnson, C.A. and Isabelle, E., Peripheral motion detection and refractive error. Science, 1972, 177, 1207-1208.

Leibowitz, H.W., Rodemer, C.S. and Dichgans, J., The independence of dynamic spatial orientation from luminance and refractive error. Perception and Psychophysics, 1979, 25, 75-79.

Lennie, P., Parallel visual pathways: a review. Vision Research, 1980, 20, 561-594.

Levinson, E. and Sekuler, R., Spatio-temporal contrast sensitivities for moving and flickering stimuli. Journal of the Optical Society of America, 1973, 63, 1296.

Levinson, E. and Sekuler, R., The independence of channels in human vision selective for direction of movement. Journal of Physiology, 1975, 250, 347-366.

MacLeod, D.I.A., Visual sensitivity. Annual Review of Psychology, 1978, 29, 613-645.

Mates, B., Effect of reference marks and luminance on discrimination of movement. Journal of Psychology, 1969, 73, 209-221.

Mates, B. and Graham, C.H., Effects of rectangle length on the velocity threshold of real movement. Proceedings of the National Academy of Science, 1970, 65, 516-520.

McColgin, F., Movement thresholds in peripheral vision. Ann Arbor, Michigan, University Microfilms, Inc., 1959.

McColgin, F., Movement thresholds in peripheral vision. Journal of the Optical Society of America, 1960, 50, 774.

Monasterio, F.M. de and Gouras, P., Functional properties of ganglion cells of the rhesus monkey retina. Journal of Physiology, 1975, 251, 167-195.

Morgan, M.J., Perception of continuity in stroboscopic motion: a temporal frequency analysis. Vision Research, 1979, 19, 491-501.

Movshon, J.A., Thompson, I.D. and Tolhurst, D.J., Spatial and temporal contrast sensitivity of neurons in area 17 and 18 of the cat's visual cortex. Journal of Physiology, 1978, 283, 101-120.

Neff, W.S., A critical investigation of the visual apprehension of movement. American Journal of Psychology, 1936, 48, 1-42.

Nes, F.L. van, Experimental studies in spatio temporal contrast transfer by the human eye, Rotterdam, Bronder-offset, 1968.

Nes, F.L. van, Koenderink, J.J., Nas, H. and Bouman, M.A., Spatio temporal modulation transfer in the human eye. Journal of the Optical Society of America, 1967, 57, 1082-1088.

Neuhaus, W., Experimentelle Untersuchung der Scheinbewegung. Archiv für die Gesamte Psychologie, 1930, 75, 315-458.

Orban, G.A., Visual cortical mechanisms of movement perception, Leuven, Vender, 1975.

Orban, G.A., Area 18 of the cat: the first step in processing visual movement information. Perception, 1977, 6, 501-511.

Orban, G.A., Kennedy, A. and Maes, H., Velocity sensitivity of areas 17 and 18 of the cat. Acta Psychologica, 1981, 48, (Special issue on the Perception of Motion) 303-309.

Pollock, W.T., The visibility of a target as a function of its speed of movement. Journal of Experimental Psychology, 1953, 45, 447-454.

Ransom-Hogg, A., and Spillman, L., Perceptive field size in fovea and periphery of the light and dark adapted retina. Vision Research, 1980, 20, 221-228.

Richards, W., Motion perception in man and other animals. Brain Behavior and Evolution, 1971, 4, 162-181.

Rijsdijk, J.K., Kroon, J.N. and Wildt, G.J. van der, Contrast sensitivity as a function of position on the retina. Vision Research, 1980, 20, 235-241.

Robson, J.G., Spatial and temporal contrast sensitivity of the visual system. Journal of the Optical Society of America, 1966, 56, 1141-1142.

Rovamo, J., Virsu, V. and Nasanen, R., Cortical magnification factor predicts the photopic contrast sensitivity of peripheral vision. Nature, 1978, 271, 55-56.

Rovamo, J. and Virsu, V., An estimation and application of the human cortical magnification factor. Experimental Brain Research, 1979, 37, 495-510.

Salvatore, S., Spatial summation in motion perception. In: J.C. Armington, J. Krauskopf and B.R. Wooten (Eds.). Visual

Psychophysics and Physiology, 1978, New York, Academic Press.

Saucer, R.T., The nature of perceptual processes. Science, 1953, 117, 556-558.

Saucer, R.T., Processes of motion perception, Science, 1954, 120, 806-807.

Schneider, G.E., Contrasting visuomotor functions of tectum and cortex in the folden Hamster. Psychologische Forschung, 1968, 31, 52-62.

Sekuler, R., Visual motion perception. In: E.C. Carterette and H.P. Friedman (Eds.). Handbook of Perception, Vol. V. Seeing. 1975, New York, Academic Press.

Sekuler, R. and Levinson T., Mechanisms of motion perception. Psychologia, 1974, 17, 38-49.

Sekuler, R., Pantle, A. and Levinson, L., Physiological basis of motion perception. In: R. Held, H.W. Leibowitz and H.L. Teuber (Eds.). Handbook of Sensory Physiology. Vol. VIII. Perception. 1978, Berlin, Springer-Verlag.

Sgro, F.J., Beta motion thresholds. Journal of Experimental Psychology, 1963, 66, 281-285.

Sharpe, C.R., The contrast sensitivity of the peripheral visual field to drifting sinusoidal gratings. Vision Research, 1974, 14, 905-906.

Sharpe, C.R. and Tolhurst, D.J., The effects of temporal modulation on the orientation channels of the human visual system. Perception, 1973, 2, 23-29.

Tolhurst, D.J., Separate channels for the analysis of the shape and the movement of a moving visual stimulus. Journal of Physiology, 1973, 231, 385-402.

Tolhurst, D.J., Sharpe, C.R. and Hart, G., The analysis of the drift rate of moving sinusoidal gratings. Vision Research, 1973, 12, 2545-2555.

Trevarthen, C.B., Two mechanisms of vision in primates. Psychologische Forschung, 1968, 31, 299-337.

Tulunay-Keesey, U., Effects of fixational eye movements on contrast sensitivity. In: J.C. Armington, J. Krauskopf and B.R. Wooten (Eds.). Visual Psychophysics and Physiology, 1978. New York, Academic Press.

Tulunay-Keesey, U. and Jones, R.M., The effect of micromovements of the eye and exposure duration on contrast sensitivity. Vision Research, 1976, 16, 481-488.

Tyler, C.W., Temporal characteristics in apparent movement: Omega movement vs. phi movement. Quarterly Journal of Experimental Psychology, 1973, 25, 182-192.

Tyler, C.W. and Torres, J., Frequency response characteristics for sinusoidal movement in the fovea and periphery. Perception and Psychophysics, 1972, 12, 232-236.

Virsu, V. and Rovamo, J., Visual resolution contrast sensitivity and the cortical magnification factor. Experimental Brain Research, 1979, 37, 475-494.

Warden, C.J., Brown, H.C. and Ross, S., A study of individual differences in motion study at scotopic levels of illumination. Journal of Experimental Psychology, 1945, 35, 57-70.

Watanabe, A., Mori, T., Nagata, S. and Hiwatashi, K., Spatial sine-wave response of the human visual system. Vision Research, 1968, 8, 1245-1263.

Wertheimer, M., Experimentelle Studien uber das Sehen von Bewegung. Zeitschrift fur Psychologie, 1912, 61, 161-265. Translated in part in Classics in Psychology, 1961. T. Shipley (ed.), Philosophical Library, New York.

Wildt, G.J. van der, Keemink, C.J. and Brink, G. van den, Gradient detection and contrast transfer by the human eye. Vision Research, 1976, 16, 1047-1053.

Wilson, H.R., Spatio-temporal characterization of a transient mechanism in the human visual system. Vision Research, 1980, 20, 443-452.

Wilson, J.R. and Sherman, S.M., Receptive-field characteristics of neurons in cat striate cortex: Changes with visual field eccentricity. Journal of Neurophysiology, 1976, 39, 512-533.

Zeeman, W.P.C. and Roeloffs, C.O., Some aspects of apparent motion. Acta Psychologica, 1953, 9, 158-181.

PSYCHOPHYSICS OF MOTION PERCEPTION*

Robert Sekuler, Karlene Ball,
Paul Tynan and Joan Machamer

Cresap Neuroscience Laboratory
Dept. of Psychology, Northwestern University
Evanston, Illinois, United States

SUMMARY

In this paper we make three major points about motion perception. As often happens, these same three points are echoed by other contributions to this Symposium. So we begin by using the ideas and words of others to anticipate what we ourselves wish to say. First, Berkley's contribution (1981) reminds readers that motion perception is a complex skill, dependent on the combination of many abilities. Starting from a quite different perspective, Bonnet (1981) arrived independently at a similar position: what is called "motion perception" is actually a heterogeneous collection of diverse functions, not a single monolithic response. The papers by Berkley and Bonnet remind us how important it is to approach motion perception through a variety of complimentary paradigms, procedures and techniques. In fact, Bonnet has provided us with an excellent taxonomy of the paradigms used in the psychophysical study of motion perception. The reader of our paper, or any paper in the Symposium, will certainly profit from Bonnet's thoughtful taxonomy and exhaustive bibliography.
    Several symposiasts demonstrated how inhomogeneous motion responses are throughout the visual field. For example, Orban (1981) showed the properties of single neurons in the cat's visual cortex vary with the spatial location of their receptive fields. Discussing applied aspects of motion perception, Leibowitz et al. (1981)

---

* We would like to acknowledge the research support provided by the National Science Foundation and the Air Force Office of Scientific Research.

emphasized differences between responses to moving targets mediated
by central and peripheral vision. The second lesson then is that if
we are to understand motion perception in its entirety, we must
examine it at different points in the visual field. Unfortunately,
we cannot assume that measurements made in one part of the visual
field can be generalized to other parts.

A third lesson we learned from Pasternak et al. (1981) and
Bonnet (1981). Although one can focus, as we shall in this paper,
on responses to <u>direction</u> of motion, one cannot ignore another
important variable, velocity. Along these lines Pasternak reported
that the difference threshold for direction of motion depends upon
the target speed used to measure it. In one of the main points of
his contribution, Bónnet argued that motion perception undergoes
qualitative changes as target speed varies.

These then are the three major points we wish to reinforce in
our paper. The nature of motion perception itself requires that
1) motion be approached with a battery of paradigms, 2) responses
in different parts of the visual field be compared, and (3) measurements be made at as large a range of target speeds as possible.

Now we can turn to the plan we shall follow in expressing
these three ideas about motion perception. First, we shall describe
the stimuli and general methods used in the experimental work. Then
we shall outline the model that guides our work. Next will come a
consideration of the model's details, particularly as they bear
upon the character of visual mechanisms enabling us to see moving
targets.

The concluding section of the paper takes us further from the
laboratory, to the consequences for motion perception of an observer's inability to predict precisely what moving target he is looking for. This sort of effect, called <u>stimulus uncertainty</u>, has
occupied a good deal of our time in the past three years and has
taught us much about motion perception and about perception more
generally. In the brief treatment of stimulus uncertainty here, we
will consider three separate but complementary topics: performance
losses associated with stimulus uncertainty, how those losses can
be compensated for or mitigated, and the discrepancy between being
able to see a target is moving and being able to identify the
direction in which it moves.

1. STIMULI AND DEPENDENT MEASURES

The stimuli used in all the work we shall discuss were patterns
of isotropic, random dot patterns presented as luminance increments
on a cathode ray tube (CRT). The spatial, temporal and intensive
properties of the dot patterns were controlled in real time by a
small laboratory computer. Their isotropy means the dot patterns
had equal energy along all axes. This property is important since
we wished to study responses to movement in as pure a form as
possible, without the complications that would be introduced by the

presence of oriented contours. The isotropy of our patterns was assessed in two ways: by numerical two-dimensional Fourier analysis and by visual inspection of the pattern's optical transform (Lipson, 1972).

Typically, the CRT on which dot patterns were displayed was illuminated by a constant veiling light of slightly less than 2 cd/m^2. In some studies, the incremental luminance of the dot patterns was an independent variable; in others it was fixed at one suprathreshold level, usually a level 50 times detection threshold. Observers sat with heads supported in a chin rest some 57 cm from the CRT. Viewing was usually binocular (we shall indicate when we come to the two exceptions to this rule). In addition, the CRT was most often masked with a circular aperture of 8 degrees diameter. The surrounding 15 x 15 degree area was illuminated to approximately the same level as the display region. In any one frame of the display (33 msec), the computer plotted 512 dots on the CRT; typically, slightly more than 400 dots were visible within the 8 degree aperture at any one time.

The dots within a pattern moved along parallel tracks. Put another way, as the dot patterns moved, dots remained in the same relative spatial phases with respect to one another. Opposite sides of the display were functionally connected so that a dot moving off one side would reappear a moment later on the opposite side. This connection gave the display the appearance of an infinite, textured surface moving continuously behind the aperture.

Our experiments have made use of several different indicator responses and trial structures. This variety of approaches is necessary to obtain the converging operations that we must have to be sure any particular description of motion sensitivity is not simply a singular outcome of one method.

We shall outline three of the major procedures we have used: reactiom time (RT) to motion onset, two-alternative forced choice (2-AFC) detection of motion and a rating-scale procedure derived from signal detection theory. This last method allows us to measure an observer's bias as well as his sensitivity to motion.

RT to Motion Onset. In this procedure, the incremental luminance of the dot patterns was adjusted to make them easily visible. Thus we were not concerned with detectability in the usual sense. On each trial dot patterns appeared first as a stationary pattern. Then, after a random foreperiod ranging from 2 to 3.5 seconds, the dot patterns began moving without warning to the observer.

The observer pressed a switch as soon as he detected the initiation of the movement. It is important to emphasize that the observer did not have to judge the properties of the movement, e.g. direction or speed. He only had to press a key at motion onset. As soon as the key was pressed the pattern disappeared and the CRT remained blank until the next trial. The dependent measure was the time between motion onset and the observer's key press (i.e., the RT).

2-AFC Testing. Each trial consisted of two, 660 msec intervals,

defined by co-extensive, high pitched tones. During one interval
the CRT was blank; during the other moving dot patterns were presented on the CRT. The interval, first or second, containing the
moving dot patterns varied randomly from one trial to the next. The
observer's task was to identify the interval containing motion. The
dependent measure is the percent of correct identifications. Usually,
we converted percent correct to a corresponding $\underline{d}'$ value. The
rationale was that although percent correct is not linearly related
to an observer's sensitivity, $\underline{d}'$ is.

    Rating-Scale Method. This was a variant of the yes-no rating
scale procedure used in signal detection work (McNicol, 1972). Each
trial was defined by a high pitched tone. On half the trials the
CRT was blank; on remaining trials moving dot patterns were presented. Depending upon the experimental conditions, dots were presented
for durations ranging from 66 to 1000 msec. In most of the work
described here, the duration was 500 msec.

    After each trial, the observer used a rating scale (the numbers
from 1-6) to describe his judgment about whether or not motion had
been presented and his confidence in that judgment. These ratings
are conditionalized upon their stimuli, cumulated and converted
using standard procedures into two non-parametric statistics: $P(\underline{A})$,
a measure of sensitivity, and $\underline{B}$, a measure of the observer's criterion. $P(\underline{A})$ is the decimal fraction of the unit square's area lying
below the receiver operating characteristic. $\underline{B}$ is analogous to
"beta" in parametric treatments of signal detection. We transform
$P(\underline{A})$ to $zP(\underline{A})$ to obtain a measure linear with sensitivity or $\underline{d}'$.
The availability of the dual measures allows us to separate sensi-
tivity changes from those associated with criterion or motivation.
The full rationale behind the rating scale method and the calculations we have used can be found in McNicol (1972).

2. ELEMENTS OF A MODEL

    Our main theoretical concern is with mechanisms best described
as "directionally-selective". These mechanisms can be treated as
filters, attenuating some input signals more strongly than others.
The input signals we have in mind are stimuli moving in one direction or another. As with other filters, a directionally-selective
mechanism can be characterized in terms of its optimum input (here,
its center direction) and the rate at which its response changes as
inputs diverge from that optimum input. This rate of change in
response is usually referred to as sharpness of tuning.

    Figure 1 shows a set of directionally-selective mechanisms
arrayed along the direction continuum. For convenience we have
given the mechanisms triangular sensitivity profiles - though much
of the model's behavior would be unaltered by substituting other
functions that decline symmetrically and monotonically with distance from the center direction. We have settled on functions that
are symmetric and monotonically declining because of selective-

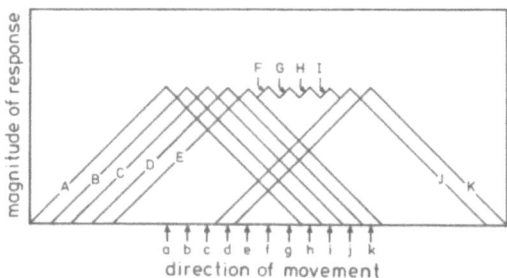

Fig. 1. A hypothetical set of sensitivity profiles for directionally-selective mechanisms, A-K, arrayed along a portion of
the direction continuum. The magnitude of response evoked
in any mechanism by some stimulus is represented by the
height of the mechanism at the point corresponding to the
stimulus direction. Mechanism A is most sensitive to direction a, mechanism B is most sensitive to direction b, and
so on. Note however, that mechanism A is also sensitive
to movement in directions b, c, d, e, f, and g, though to
a reduced degree. Mechanism A is not sensitive to movement
in directions h-k, so movement in those directions evokes
no response in A.

adaptation experiments with random dot patterns. Levinson and
Sekuler (1980) used that procedure to produce a two-dimensional
picture of selectivity for direction of motion. After observers
adapted to a pattern of dot patterns moving in one direction, the
luminance detection threshold for patterns of test dot patterns
moving in the same or similar directions was elevated. As adapting
and test directions diverged, threshold elevation produced by the
adapting pattern decreased.

The "tuning curves" for directional selectivity measured by

this technique were rather broad. In fact, some elevation is present even when test and adapting directions differ by 45 degrees. By the time the directions differ by 70 degrees or so, their interactions have dropped to zero. We wish to emphasize that these effects are directional, not axial in character. First, there is no threshold elevation – nor facilitation, for that matter – when test and adapting directions differ by 180 degrees. Second, the tuning disappears when observers use an alternative criterion, one related to the detection of pattern rather than motion per se. Further details are provided in Levinson and Sekuler (1980).

Other data, too, demonstrate that the tuning curves of directionally-selective mechanisms are symmetric and monotonic. Here we shall consider only one such demonstration, derived from the noise masking experiments of Ball and Sekuler (1979). They required observers to detect moving dot patterns following exposure to noise containing various directional components. Here, two main types of noise are of interest. One kind can be characterized as "broadband" – the noise contained equal amounts of all directions of motion (consider the analogy to white noise). The other kind was noise from which certain bands of directions had been digitally filtered. Varying the band of noise filtered from the noise, Ball and Sekuler examined how various direction bands affected the detection of some test direction.

Consider the rationale for this approach. Suppose that the detection of some test direction depended upon the response of a particular directionally-selective mechanism. Depending upon its directional content, the masking stimulus could inject noise into the filter whose response signals the presence of the test direction. As would be true for any real system, with temporal impulse response of greater than zero duration, the effect of the visual noise would outlast the masking stimulus, lowering the effective signal/noise ratio produced by the subsequently presented dot patterns moving in one direction.

The dependent measure was RT to the onset of unidirectional motion immediately following the masking noise. Since the masking noise was of variable duration (2-3.5 sec, akin to a random foreperiod in the usual RT experiment) the observer could not divine – without actually seeing the unidirectional test motion – when the motion would occur. In this way, RT to motion onset could be used as an index of the visibility of the motion.

For any one test direction, RT is elevated by those components of noise that are most nearly in the same direction as the test. When those similar components are filtered out of the noise, the noise is unable to elevate RT. As components are filtered out, RT declines symmetrically and monotonically with respect to the test direction. Again, the tuning is broad; noise components as different from the test direction as 45 degrees significantly elevate RT to the test direction but the effect of noise components 70 degrees from the test direction is virtually nil.

Returning to Fig. 1, consider another feature of the direc-

tionally-selective mechanisms portrayed there. The magnitude of
response evoked in any mechanism by motion of constant speed and
contrast is represented by the <u>height</u> of the tuning function at the
point on the direction continuum that corresponds to the stimulus
direction of interest. Note that all mechanisms are shown with the
same sensitivity at their center directions. The assumption of
equal peak sensitivities derives from several demonstrations that
sensitivity to moving targets is independent of their direction of
motion. Let us briefly consider the evidence for this assumption.

Recent work has provided three separate demonstrations relevant
to this question. First, Ball and Sekuler (1980) obtained RTs to
the movement of dot patterns in various directions. These RTs,
measured in the absence of masking noise, were invariant with direction. Second, Levinson and Sekuler (1980) measured contrast thresholds for moving dot patterns. These, too, were independent of direction. Finally, Marshak (1981) measured the duration of motion after-
effects (Waterfall Illusion) produced by exposure to dot patterns
moving in various directions and found that after-effect duration
was constant for all directions tested. The results of all three
studies are summarized in Figure 2: using three different measures
of response to motion all demonstrate that sensitivity to motion
does not vary with direction. Hence, all the tuning functions in
Fig. 1 should have the same height.

Returning once again to Fig. 1, note that the directionally-
selective mechanisms are represented as having uniform breadth of
tuning. While this may be nearly correct within restricted regions
of the direction continuum, it most certainly is not correct if one
tries to describe the entire continuum. There are several converging
demonstrations that tuning varies with center direction. The ones
we shall discuss compare tuning about an oblique direction (45 degrees) to tuning about a direction along a major axis (90 degrees).
Let us consider each of these demonstrations in turn.

Using the noise masking paradigm mentioned earlier, Ball and
Sekuler (1980) found that tuning functions were broader when the
test direction was 45 degrees than when it was 90 degrees. Within
the logic of their approach, this difference implies that the directionally-selective mechanism responsible for detecting upward
motion (90 degrees) is more narrowly tuned than the comparable
mechanism responsible for detecting motion in an oblique direction.
Since their original dependent measure was RT to motion onset, Ball
and Sekuler wanted to verify the conclusion using a procedure more
directly related to tuning in a conventional sense. They reasoned
that these differences in tuning should also show up in a differential ability to detect changes in direction of motion around
90 degrees and changes in direction around 45 degrees.

As a test, they used the method of constant stimuli to measure
direction difference thresholds with two different standard directions, 90 and 45 degrees. Each trial consisted of two intervals,
600 msec long, separated by an interval of one second. The first
interval contained motion in the standard direction; the second

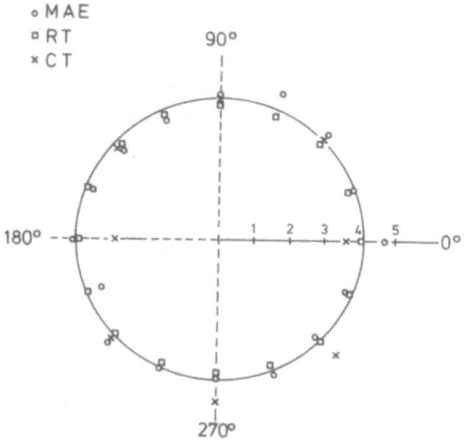

Fig. 2. Three different psychophysical responses to motion as a
    function of direction. Circles represent the duration of
    motion after-effect (MAE) elicited by various directions;
    Squares represent the reaction time to onset of motion in
    various directions (RT); "x"s represent the contrast thres-
    hold for seeing dot patterns that move in various direc-
    tions (CT). All three measures have been scaled so that
    their means coincide.

interval contained motion in a comparison direction – either the
same as the standard, 1 degree or 2 degrees clockwise or counter-
clockwise relative to the standard. The observer judged the direction of movement in the second interval relative to that in the
first. Difference thresholds were obtained from the least squares
fits to the psychometric functions. The difference threshold for
upward motion was 3.45 degrees and that for oblique motion was
8.42 degrees. A similar result has been obtained by Machamer (unpublished) as part of a study of direction difference thresholds
for dot patterns moving at several speeds and in various directions.
She, too, found the difference threshold for 90 degrees to be about-
half that for 45 degrees.
    These results with difference thresholds are consistent with

differential tuning for mechanisms sensitive to oblique and upward
motion. The following formalizes the necessary argument. Consider
a stimulus of a direction appropriate to produce a maximal response
from one mechanism in Fig. 1 (the mechanism's center direction). The
response of the mechanism is given by the product of a) the intensity of the stimulus and b) the sensitivity of the mechanism to that
stimulus. Obviously, if the direction of the stimulus changes, the
mechanism's response will decrease. Assume that a just noticeable
change in direction requires a criterion change in the mechanism's
response. A broadly tuned mechanism will yield that criterion change
only after a larger stimulus change than would be required to produce the same criterion change in a more sharply tuned mechanism.

3. EXTENSIONS TO THE SIMPLE MODEL

The model just sketched fails to provide all the detail we
would like about motion perception, even in the simplest situations.
For example, target speed is ignored. One justification for this
particular omission is that speed and direction are probably coded
independently by human visual mechanisms (Ball and Sekuler, 1980).
But, both psychophysics and physiology suggest that the neural coding
of target velocity presents an intriguing problem all by itself.
Obviously, no model of motion perception can claim completeness if
it ignored the problem of perceived speed. At various levels of the
mammalian visual system, the receptive fields of cells that respond
to moderate and high rates of temporal modulation tend to be more
uniformly distributed across the visual field than are those of
cells that are less responsive to such temporal modulation (Fukuda
and Stone, 1974; Kirk, Levick and Cleland, 1976). This difference
between retinotopic distributions led Tynan and Sekuler (in press)
to seek corresponding differences in retinotopic distribution of
psychophysical responses.
They wondered whether the distributions of receptive fields of
cells responsive to different rates of temporal modulation might
affect psychophysical responses to moving targets at various eccentricities. This led them to examine two dependent variables at
various retinal eccentricities. Based on the physiological data
above, their hypothesis was that, with sufficiently high target
speeds (and correspondingly high rates of temporal modulation)
psychophysical responses would be invariant with eccentricity.
In their first experiment, Tynan and Sekuler measured RT to
motion onset for upward moving dot patterns presented at various
eccentricities. The screen of the CRT was masked by a 10 degree
diameter circular aperture. An electronic blanking circuit eliminated dot patterns from either the center of the screen or from its
periphery. As a result, the circuit produced either a <u>patch</u> of dot
patterns in the middle of the screen, or a central area devoid of
dot patterns surrounded by an <u>annulus</u> of dot patterns. With either
<u>central patch</u> or <u>annulus</u>, dots moving into the blanked zone dis-

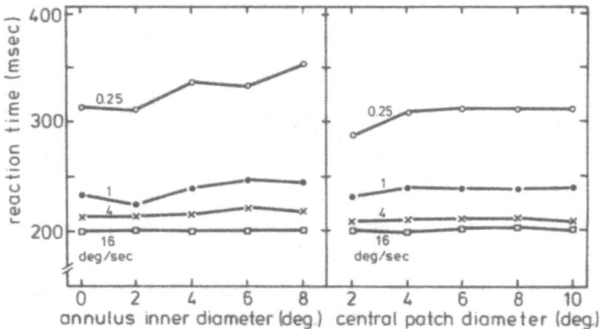

Fig. 3. Reaction times to dot patterns moving at various velocities.
Left panel shows results using annular stimuli of various
diameters right panel shows results with stimuli restricted
to central patches. See text for details.

appeared; dots leaving the blanked zone reappeared.

Annuli were either 0, 2, 4, 6 or 8 degrees in inner diameter;
central stimulus patches were either 2, 4, 6, 8 or 10 degrees in
diameter. RTs were measured to stimulus velocities of 0.25, 1, 4,
and 16 deg/sec. The main results of this experiment are shown in
Fig. 3. With annular stimuli (left panel), the lowest velocity,
0.25 deg/sec, yielded RTs that increased steadily with annulus size.
At higher velocities, RT was independent of annulus size. With
central patches of moving dot patterns (right panel), patch size
influenced RT only between a 2 degree patch and one of 4 degrees at
the lowest speed used. For all higher velocities, RTs were invariant
with patch size. Note, in addition, that in both panels, RT declines with increasing stimulus velocity.

Visual functions that depend upon spatial resolution - acuity
and, very likely, RT to very slow movement - fall off dramatically
over the portion of the field studied in this experiment (LeGrand,
1967). We believe that such visual functions depend upon physiolo-

gical mechanisms that respond preferentially to lower temporal frequencies. The rapid decline in psychophysical spatial resolution is consistent with the hypothesis that cells responsive to lower temporal frequencies are more likely to have receptive fields in the center of vision. RTs to moderate speeds of motion show no decline over this same range of eccentricities. Very likely, cells with appreciable sensitivity to higher rates of temporal modulation participate in the detection of such motion. The invariance in RT with eccentricity is consistent with the hypothesis that cells with receptive fields in the periphery respond to higher temporal rates.

In a second, related experiment, Tynan and Sekuler measured the perceived speed of targets at various eccentricities. These measurements were made for targets covering a range of speeds. Stimuli were random dot patterns moving upward within a strip 28 degrees high by 4.7 degrees wide. Observers matched the apparent speed of a target at each of several eccentricities with the adjustable speed of similar dot patterns presented in the center of vision. Test targets could be presented immediately to the left of the fixation point, or at various distances from it: 7.5, 15, 22.5, and 30 degrees. The duration of any movement varied randomly between 1.5 and 2.5 seconds, making it impossible to judge velocity simply from the distance traveled by any particular element in the pattern. Target velocities of 0.25, 1, 4, and 16 deg/sec were factorially combined with the five eccentricities.

Fig. 4 shows the mean velocity matches to dot patterns moving at 0.25, 1, 4, and 16 deg/sec. Note that no data are given for the 0.25 deg/sec stimulus at eccentricities beyond 7.5 degrees. These data have been omitted because on more than half the trials with such eccentricities, the 0.25 deg/sec stimulus appeared stationary, a phenomenon reported by Lichtenstein some years ago (1963). Values plotted against the ordinate have been normalized by dividing each by the actual speed of the test pattern. Plotted in this way, ordinate values less than unity indicate the eccentric target appeared to move more slowly than it would have with central viewing. The results of the experiments can be summarized simply: eccentrically-viewed dot patterns appear to move more slowly than do centrally-viewed ones. This slowing effect increases with eccentricity and decreases with target speed.

In both these experiments, psychophysical responses to slowly moving targets change rapidly as a function of eccentricity of presentation. Also, in both cases, psychophysical responses to rapidly moving targets are nearly invariant with eccentricity of presentation. These experiments sought to test an hypothesis about psychophysical parallels to the retinopic distributions of neural cells whose temporal responses differ from one another. The effects obtained seem to parallel the retinotopic distributions of neural cells that respond best to low rates of temporal modulation and of neural cells that respond best to higher rates of modulation.

Obviously, there needs to be a follow-up with other kinds of temporally modulated stimuli at various eccentricities. Such stimuli

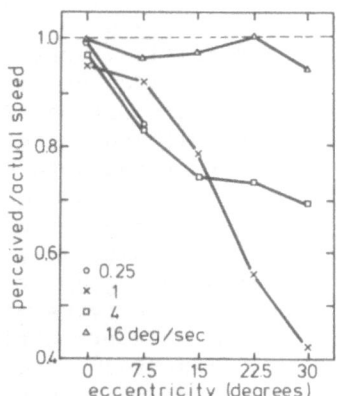

Fig. 4. Ratio of perceived to actual speed as a function of target
        eccentricity. Data are shown for several different target
        speeds, 0.25, 1, 4, and 16 deg/sec. Missing data are due
        to apparent standstill at certain combinations of speed and
        eccentricity.

should include spatially localized targets whose eccentricity can
be specified more precisely.

     But, follow-ups aside, the work just described does offer an
important lesson that others should consider. Although many psychophysical theorists have found it convenient to dichotomize visual
mechanisms into "sustained" and "transient", the visual system very
likely does not itself always respect this bipartite classification.
But it is easy to see how one could be misled into believing such
a dichotomy characterized the structure. For example, if Tynan and
Sekuler had considered only extreme speeds, there would have been a
clear difference in visual response as a function of eccentricity.
This clear separation simulated a dichotomy. However, if we also
take account of intermediate speeds, responses define a continuum
between these extremes. For such intermediate velocities, psychophysical responses are neither invariant with eccentricity nor do
they exhibit as rapid a decline as do responses to the most slowly

moving targets. As we have been so well reminded by Kelly (1977), stimuli for many visual responses must be defined on a continuum of both spatial and temporal dimensions.

4. STUDIES OF STIMULUS UNCERTAINTY

When an observer cannot anticipate precisely the characteristics of a moving target that he has to detect, the visibility of that target is drastically reduced (Sekuler and Ball, 1977). Outside the laboratory, the vast bulk of our responses to moving targets come in face of such uncertainty. But what is the effect of not knowing precisely the sort of moving target one is looking for? In the laboratory, and presumably outside as well, inability to anticipate a target's precise speed or its direction of motion produces a substantial performance loss. Such losses in detectability have been demonstrated using a variety of psychophysical procedures, including criterion-free ones. Since our group at Northwestern has published extensively on the effects of uncertainty, we will not repeat those findings here. Instead, let us describe some new work on certain aspects of direction uncertainty that may illuminate motion perception in general.

The first experiment is one of a series dealing with the amelioration of performance losses from uncertainty. The strategy used by Ball and Sekuler (in press) was to provide cues to the observer that told him with varying degrees of reliability what direction he would have to detect. They asked how reliable the cue had to be in order to diminish the effect of direction uncertainty. For example, did the cue have to be exactly right in order to help the observer? To answer the question, three types of blocks of trials were run. In blocks of Certainty trials, all movement was in the same direction. Thus, there was no direction uncertainty. In blocks of Uncued-Uncertainty, the direction in which the dot pattern moved was drawn randomly from a uniform distribution covering all 360 degrees and no cue was provided. In blocks of Cued-Uncertainty, an oriented line appeared 700 msec before the onset of motion to provide the observer a cue as to direction of motion that would occur. Each trial consisted of a warning signal, the cue (oriented line on for 50 msec), the 700 msec delay, and finally the observation interval of 500 msec.

Seven different levels of cue reliability were run (360, 300, 240, 180, 120, 60, and 0 degrees). To understand the meaning of cue reliability, consider three examples. In the 360 degree condition the cue orientation was perfectly unreliable; regardless of the cue orientation, subsequent movement could be in any direction. In the 180 degree condition, stimulus movement following the cue was randomly chosen from a uniform distribution within 180 degrees of the indicated direction. Thus the motion could vary by 90 degrees either side of the direction indicated by the cue. In the 0 degree condition, the direction of movement was identical to that indicated

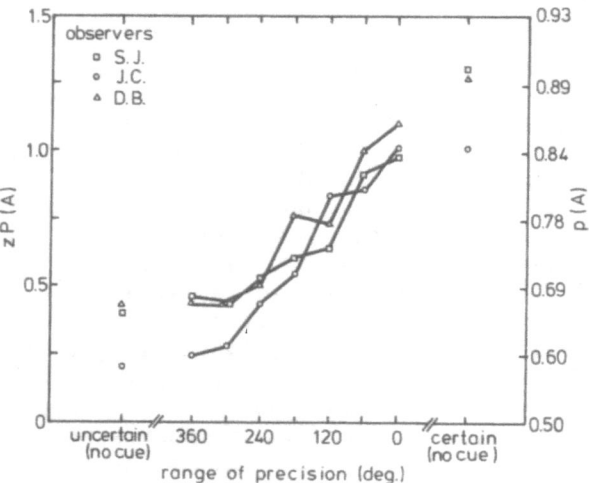

Fig. 5. Sensitivity zP(A), to moving dot patterns. Measurements are
        shown for three observers. Middle section of x-axis,
        labelled "range of precision", shows the effect of provid-
        ing cues, before the motion occurs. The reliabilit of the
        cue increases from left to right and the effect of direc-
        tion uncertainty decreases. Data points at extreme left
        show performance with complete direction uncertainty and
        no cue to help the observer. Data points at the extreme
        right show performance with no direction uncertainty (all
        patterns move in the same direction).

by the cue. This condition provides a perfectly reliable cue. Movement was presented on half of the 80 trials per block. On the other
half, the CRT was illuminated only by the steady veiling light.
Observers judged that a moving pattern had or had not been presented using the rating scale described earlier. Observers were informed before each block of trials which condition was to be presented.
    The results are shown in Fig. 5. Sensitivity measures are
plotted against various cue reliabilities. The three least reliable
cue conditions (360, 300, and 240 degrees) were significantly worse
than all other conditions (p<.05). The 180 and 120 degree range
conditions did not differ significantly from each other, but were

significantly worse than the 60 or 0 degree conditions ($\underline{p}$<.05).
Similarly, 60 and 0 degree conditions did not differ from each other
but were significantly lower than the certainty condition. Thus as
the cue became more and more reliable, sensitivity increased and the
detrimental effects of uncertainty declined.

To determine if these results were caused by a variation in
criterion - willingness or reluctance to guess that movement had
occurred - $\underline{B}$ values were also calculated. An analysis of variance
showed no significant change across conditions indicating that the
observer's bias for or against saying that motion had been presented
did not vary across conditions.

A very reliable cue delivered 700 msec in advance of the target
virtually eliminates the effect of direction uncertainty. More im-
portantly, in order to reduce the effects of direction uncertainty,
a cue does not have to specify with perfect accuracy the direction
of subsequent movement. A cue that indicated a direction of movement
only within a range of 120 degrees was still sufficient to reduce
the uncertainty effect by half.

These results have clear implications for reducing uncertainty
in non-laboratory situations. If crude but timely cues aid performance, then it will be possible to develop special hardware that
could assist human observers by doing a quick, rough analysis of a
signal and giving the observer information to reduce his uncertainty
about the target. Such a sensory cue, reduced to hardware form,
could be useful to pilots, drivers, radar operators, or anyone for
whom uncertainty about moving targets is costly. If devices of this
kind are to be helpful, however, they must deliver the information
in time to allow the observer to assimilate it and adjust his target
search accordingly. Feasible sensory aids would not allow the observer much advance warning, so it is essential to know how far in
advance this information must be provided to facilitate performance.
Furthermore, feasible sensory aids are more likely to provide only
crude information rather than specifying stimulus characteristics
precisely. But these developments are still in the planning stage.

Let us end with a quite different kind of study of direction
uncertainty, one that relates closely to the model presented earlier
(Fig. 1). In one experiment by Ball, Machamer and Sekuler (unpublished) an observer's ability to identify the direction of motion
that he had just seen was tested at various levels of visibility.
As we shall see, there were some rather striking discrepancies between detection performance - simply seeing the motion - and the
ability to assess direction of motion. Ball et al. first established
the 2-AFC detection levels for a wide range of combinations of
target duration and contrast. These data are of interest in their
own right, but are not directly relevant here. However, we shall
have to cite some of them because they provide reference points for
the identification data that are of primary interest.

A moving dot pattern was presented on every trial and the observer judged its direction. Stimuli were drawn randomly from a
uniform distribution of directions covering the range from 75 to

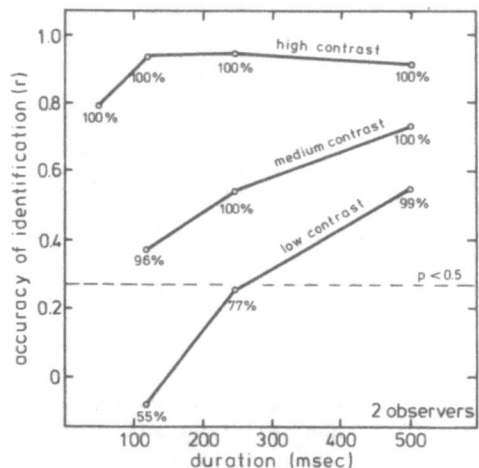

Fig. 6. Pearson product moment correlation (<u>r</u>) between reported and
actual directions of motion for patterns of various duration. The parameter of the family of curves is the contrast
of the dot patterns; percentages indicate the 2-AFC detection performance for each of the data points. Dashed
horizontal line indicates <u>r</u> value at which the correlation
between reported and actual directions is significant
(<u>p</u><.05).

105 degrees. A protractor around the display aided the observer in
reporting perceived direction. A trial consisted of a variable
duration moving pattern following which the observer reported the
perceived direction of motion. In a block of trials, each of the
31 possible directions appeared three times in random order.

Ball et al. calculated a) the correlation between the direction actually presented and the observer's judgment, as well as
b) the observer's average error in judgment. The average error was
based upon the absolute value of the difference between the actual

and perceived directions. Some of these results are shown in Fig. 6
where 2-AFC detection performance for each stimulus is indicated by
the number above each data point. Significant correlations (p<.05)
lie above the dotted horizontal line.

First, note identification performance when the random dot
display was presented at the highest contrast for the longest durations, 500 msec. Correlations between perceived and actual direction
of movement averaged 0.93 and 0.89 for the two observers. So, under
optimal conditions observers can report direction of movement with
a fair degree of accuracy. The average errors (2-4 degrees) correspond well to direction difference thresholds measured under comparable conditions. At this high contrast observers detected movement 100% of the time with virtually no strain. The middle curve
shows identification performance at a lower contrast value - but one
still permitting 100% or near 100% detection. At this contrast, ability to report the direction of movement falls off as duration is
shortened. Oddly enough, contrast and duration must be high enough
to mediate nearly 100% detection if the observer's judgment of direction is to be better than chance. The results demonstrate that detection of moving dot patterns may sometimes be far easier than
judgment of their direction,

These findings were extended in other detection-cum-identifica-
tion experiments by Ball et al. One of those experiments gives
another estimate of the breadth of tuning of directionally-selective
mechanisms. The procedure was a 2x2-AFC experiment. Each trial consisted of two intervals; moving dots were presented in one, nothing
in the other. The motion that did occur was randomly drawn from
pairs of alternative directions. The observer's first task (detection) was to indicate the interval that contained motion. The observer's second task (identification) was to indicate which of the
two directions had been presented. Ball et al. compared detection
and identification levels for various differences between the two
alternative directions. As the difference increased, detection fell
and identification improved.

When the difference between directions had reached approximately 150 degrees, the two measures were identical and they remained so
for larger differences. The logic of Watson and Robson (1980) suggests that 150 degrees would be the minimum difference between
center directions (cf. Fig. 1) that produces no overlap between the
two mechanisms whose center directions we are considering. This
corresponds to a range of 75 degrees from the center direction of
some mechanism to the point at which its sensitivity has declined

essentially to zero*. This value is consistent with several estimates
of directional selectivity cited earlier. It seems that operations
of widely differing kinds converge on approximately this value.

These experiments suggest another point that is quite important
from the perspective of this Symposium. Here we mean the discrepancy
between ease of seeing a moving stimulus and the possibly severe
difficulties in correctly assessing its direction. Military standards specify the desired visibility for various tasks in terms of
detection levels, not in terms of identification levels. The discrepancy between the two measures suggests that since those who
create such standards are often really interested in insuring a
certain level of identifiability, the standards must be defined in
such terms. One cannot guarantee good identifiability simply by
arranging conditions to produce good detectability. The two need
not go hand in hand.

---

*For completeness we note a claim, which if it were true, would
require some modification of the model we have been describing.
Our model considers mechanisms sensitive to widely different directions of motion to act independently of one another. This assumption is grounded in much empirical evidence (see Ball and Sekuler,
1980).Moulden and Mather (1978) quarreled with this idea, suggesting that mechanisms sensitive to opposite directions of motion are
not independent of one another. They based their argument on the
claim that adaptation to one direction of motion actually facilitated detection of the opposite direction. Unfortunately, Moulden
and Mather were unable to make direct, empirical observation of
facilitation and had to rely upon tenuous, statistical "adjustments" to their actual results. They contrast their ratio theory,
in which opposite directions stand in a special relationship to
one another, to the model we present here. Moulden and Mather
"maintain that a single principle, namely that of a 'ratio' mechanism, can provide an account of both threshold and supra-threshold
phenomena (1978, p. 519)". Given the vehemence with which Moulden
and Mather championed the ratio theory in their 1978 paper one
cannot but admire the speed with which they are willing to acknowledge in print that the results of their own subsequent experiments showed the ratio theory to be in error (Mather and Moulden,
1980; Mather, 1980).

REFERENCES

Ball, K., and Sekuler, R. Masking of Motion by broadband and filter-
    ed directional noise. Perception & Psychophysics, 1979, 26,
    206-214.
Ball, K., and Sekuler, R. Models of stimulus uncertainty in motion
    perception. Psychological Review, 1980, 87, 435-469.
Ball, K., and Sekuler, R. Cues reduce direction uncertainty and
    enhance motion detection. Perception & Psychophysics, in press.
Berkley, M.A. Neural substrates of the visual perception of move-
    ment. This Volume, 1981.
Bonnet, C. Thresholds of motion perception. This Volume, 1981.
Fukuda, Y., and Stone, J. Retinal distribution and central projec-
    tions of Y-, X-, and W-cells of the cat's retina. Journal of
    Neurophysiology, 1974, 37, 749-772.
Kelly, D.H. Visual contrast sensitivity. Optica Acta, 1977, 24,
    107-129.
Kirk, D.L., Levick, W.R., and Cleland, B.G. Crossed and uncrossed
    representation of the visual field by brisk-sustained and
    brisk-transient cat retinal ganglion cells. Vision Research,
    1976, 16, 225-232.
LeGrand, Y. Light, Colour and Vision. London: Chapman and Hall,
    1968.
Leibowitz, H.W., Post, R.B., Brandt, Th., and Dichgans, J. Implica-
    tions of recent developments in dynamic spatial orientation
    and visual resolution for vehicle guidance. This Volume, 1981.
Levinson, E., and Sekuler, R. A two-dimensional analysis of direc-
    tion-specific adaptation. Vision Research, 1980, 20, 103-107.
Lichtenstein, M. Spatio-temporal factors in cessation of smooth
    apparent motion. Journal of the Optical Society of America,
    1963, 53, 302-306.
Lipson, H. Optical Transforms. New York: Academic Press, 1972.
McNicol, D. A primer of signal detection theory. London: George
    Allen and Unwin Ltd., 1972.
Marshak, W. Perceptual integration and differentiation of direc-
    tions in moving patterns. Unpublished Ph.D. dissertation,
    Northwestern University, 1981.
Mather, G. The movement aftereffect and a distribution-shift model
    for coding the direction of visual movement. Perception, 1980,
    9, 379-382.
Mather, G., and Moulden, B. A simultaneous shift in apparent direc-
    tions: Further evidence for a "distribution-shift" model of
    direction coding. Quarterly Journal of Experimental Psychology,
    1980, 32, 325-333.
Moulden, B., and Mather, G. In defence of a ratio model for move-
    ment detection at threshold. Quarterly Journal of Experimental
    Psychology, 1978, 30, 505-520.
Orban, G.A. Velocity sensitivity of areas 17 and 18 of the cat.
    Acta Psychologica, 1981, 48, (Special Issue on the Perception
    of Motion) 303-309.

Pasternak, T., Movshon, J.A., and Merigan, W.H. Motion mechanisms
    in strobe-reared cats: Psychophysical and electrophysiological
    measures. Acta Psychologica, 1981, 48, (Special Issue on the
    Perception of Motion) 321-331.
Sekuler, R. Visual motion perception. In: Handbook of Perception,
    (edited by E. Carterette and M. Friedman), Volume 5, 1975.
Sekuler, R., and Ball, K. Mental set alters visibility of moving
    targets. Science, 1977, 198, 60-62.
Tynan, P.T., and Sekuler, R. Motion processing in peripheral vision:
    Reaction time and perceived velocity. Vision Research, in
    press.
Watson, A.B., and Robson, J.G. The number of channels required for
    the identification of spatial and temporal frequencies. Paper
    presented at meetings of Association for Research in Vision
    and Ophthalmology, Sarasota, 1979.

VISUAL LOCALIZATION AND EYE MOVEMENTS*

Leonard Matin

Department of Psychology, Columbia University

New York, N.Y. 10027, U.S.A.

1. INTRODUCTION

1.1. The Fundamental Problem of Visual Localization

Displacement of the retinal image at the back of the eye may be produced either by eye movements, which the observer uses to redirect his gaze within the visual field, or by displacements of the visual field itself outside of the eye. Nevertheless while displacements of the visual field are normally seen to be displacements in the environment, stationary visual fields normally continue to appear stationary in the presence of eye movements. The main concern of this chapter is with the basis for this difference in visual localization.

Some of the technical terms that are used in the chapter are collected in the glossary immediately below. The glossary is sequential, i.e., terms defined earlier are used in later definitions.

1.2. Glossary

1. Extraretinal Signal, Extraretinal process, or Extraretinal Eye Position Information (EEPI): Any information regarding the position of his eye in the orbit that an observer has available which does not derive from stimulation of the retina by light.

---

*The writing of this chapter was supported in part by contract N 62269 80C 0296 with the Naval Air Development Center (NADC) and in part by PHS research grants EY 00375 and EY 03198 from the National Eye Institute, NIH.

2. <u>Retinal Signal, Retinal process, or Retinal Information (RI)</u>:
Information in the visual system resulting from stimulation of the
retina by light. It is not intended to indicate that the events
referred to take place at the retina, the LGN, the visual cortex,
or any other specific locus, only that they were initiated as a consequence of stimulation of the retina by light.

3. <u>Cancellation Theories</u>: A class of theories used to explain
the stability of visual localization of stationary objects in the
presence of eye movements in which RI and EEPI are subtracted from
each other; according to these theories values of the algebraic differences are simple transforms of values of perceived visual direction (see Fig. 1).

4. <u>Visual Localization</u>: Localization of visually perceived
objects relative to other visually perceived objects (either simultaneously or successively perceived), or relative to a visual norm,
or by absolute identification.

5. <u>Intersensory Localization</u>: Localization of an object sensed
by one sensory modality relative to localization of the same object
or a different object sensed by another modality; also applies when
the comparison involves a norm or norms.

6. <u>Type A Suppression</u>: Suppression of information regarding
spatial localization of a sensory stimulus by information from
another modality ("intermodal suppression"). This chapter will have
two main uses for this term (a) suppression of EEPI for use in
visual localization by the presence of a structured visual field;
(b) suppression of auditory information regarding spatial localization of an auditory stimulus by visual stimulation.

7. <u>Type B Suppression</u>: Suppression of information in a given
modality consequent on stimulation from the same modality ("intra-
modal suppression"). The particular use of Type B suppression of
greatest concern in this chapter is that in which information regarding visual localization is lost or distorted as a result of
interference by changes in visual stimulation due to eye movements.

8. <u>Type A Error of Visual Localization</u>: Error of localization
of a visual stimulus consequent on removal of Type A suppression
under conditions when EEPI is faulty.

9. <u>Type B Error of Visual localization</u>: Error of localization
of a visual stimulus due to the absence, modification, or prevention
of normal action of Type B suppression under conditions when EEPI
is faulty.

10. <u>Type A Failure of Cancellation Theory</u>: Accurate visual localization in conjunction with errors in EEPI, or inaccurate visual

localization in conjunction with accurate EEPI.

11. Type B Failure of Cancellation Theory: Failure of Cancellation Theory due to Type B error of visual localization.

12. Type A a/v Error: Error in matching the spatial location of an auditory target to a visual target in which the magnitude of the error increases linearly with gaze eccentricity; cases so far reported have been found to be due to errors in EEPI consequent on extraocular muscle weakness.

13. Type C a/v Error: Error in matching the spatial location of an auditory target to a visual target for which the magnitude of the error does not change with gaze eccentricity; cases observed so far are due to errors in auditory localization.

I have tried to reserve the "A" designation for cases in which the role of EEPI is of special concern, and the "B" designation for cases in which Type B suppression is of special concern. In this vein "Type B a/v errors" have not yet been observed. Perceptual errors due to deficits in the auditory system are neither Type A nor Type B, and the "C" designation is used for them (e.g. Type C a/v errors). Also see the comment at the end of the glossary.

14. Retinal PSE (Retinal Point of Subjective Equality): Location on the retina at which a test target must be placed in order to appear in the same visual location as another (standard) target (Retinal PSE for standard target) or a norm (Retinal PSE for norm). The Retinal PSE is a statistical measure of central tendency calculated from psychophysical reports of whether the test target lies to the left or right (above or below) of the standard target or norm. Its calculation requires that the experimenter know the location of the physical targets *and* the position of the eye in the orbit at the time of target presentation.

15. PSE (or Target PSE): Physical location of a test target that appears in the same location as another (standard) target. The Target PSE is a statistical measure of central tendency calculated from psychophysical reports of whether the test target lies to the left or right of (above or below) the standard target or norm. Its calculation requires that the experimenter knows only the location of the physical targets presented to the subject; eye position measurements are not required.

The Type A/Type B terminology for classifying varieties of suppression was introduced in Matin (1981) (its use here involves a minor terminological change from its original presentation). Type A suppression occurs with steady gaze; it is intermodal with a site of occurrence central to any peripheral sense organs; the losses are related to spatial localization. Type B suppression refers to intra-

modal suppression for which some, but not all of the sites of occurrence are peripheral; the losses occur in the presence of *changes*
in eye position; the losses are either related to detection or to
visual localization or both.

        Classifications which are not based on mutual exclusion contain
the seeds of their own demise. So with the present case where
several potential violations of exclusivity exist (e.g. visual localization as influenced by EEPI is itself influenced by visual
context). However, the separation of two types of suppression does
point to an important, and possibly fundamental, difference. It is
clearly useful at the present state of knowledge regarding spatial
localization, but it is not written in stone.

1.3. Two types of failure of cancellation theories

        Helmholtz (1866) originally proposed that stationary visual
fields appear stationary when we turn our eyes because the "effort
of will" in turning the eyes is taken into account. Since then, a
number of other theories have been put forward to explain the
stability of visual localization in the presence of eye movements
which change the location of retinal images of stationary objects.
Most of these theories (for earlier reviews see Matin, 1972,.1976a;
MacKay, 1973; Shebilske, 1977) are "Cancellation Theories" (that is,
they propose that extraretinal eye position information (EEPI) is
coupled with information regarding eye-movement-produced change in
retinal image location of stationary objects) (I use RI for Retinal
Information); the coupling is done in a mechanism whose output is
the algebraic difference of EEPI and RI. The label "Cancellation" is
applied to these theories because they suggest that the theoretical
coupling of EEPI and RI produces a nulling of the effects of one by
the other and this nulling is the correlate of perceptual stability.
Different versions of these theories differ mainly in specifying
different sources for EEPI (see Fig. 1) and in the degree of quantitative precision with which they describe a coupling operation.
        The first attempt at a systematic test of Cancellation Theory
(Matin and Pearce, 1965) uncovered the fact that although EEPI-
mediated cancellation was involved in visual localization, there are
substantial errors of visual localization of brief flashes in the
presence of voluntary saccades, and thus a failure of Cancellation
Theory. Subsequent experiments demonstrated a number of other varieties of failures in connection with perception in the presence of
voluntary saccadic eye movements (for earlier summaries, see Martin,
1972, 1976b). These include (a) large perceptual mislocalizations
of brief flashes; (b) large influences on the magnitude of the mislocalizations produced by visual backgrounds; (c) no mislocalizations in some instances in which Cancellation Theory predicts them.
        However, an even more serious failure of Cancellation Theory
has been obtained with steady fixation. This failure of theory was
not discovered until an entirely different procedure was employed

(Matin, Picoult, Stevens, Edwards, Young and MacArthur, 1980, 1981b; Matin, Stevens and Picoult, in press) which produced dramatic errors of visual localization with steady gaze. Consideration of both types of errors of visual localization leads to a theory of visual locali- zation (Dual-Suppression Theory) some of whose rudiments have been presented in several different publications (Matin, 1972, 1976b; Matin, et al., 1981b; in press). Among other things, this review will deal with both types of error, how they constitute failures of Cancellation Theory, what requirements they set for an alternative theory of perceptual stability, how Dual-Suppression Theory accounts for them, and the place of cancellation in spatial localization.

1.4. Brief statement of the main failure (type A) of cancellation theory and an important success

Type A failure of Cancellation Theory (see Glossary) was dis- covered in some experiments with partial curarization of the extra- ocular muscles (Matin et al., 1980, 1981b). It consists of the con- junction of the following three points and Conclusion 1:

(1) Large errors of EEPI are produced and are measured when the extraocular musculature is so weakened;

(2) The curare-induced error in EEPI changes systematically with the position of the eye in the orbit. For example, under one set of conditions, for one eye position (eye in steady fixation turned 10° to the right of primary position) the error in EEPI is +20°; for a second eye position (eye in steady fixation turned 10° to the left of primary position) the error in EEPI is -20°.

(3) The errors in EEPI stated in (2) are measured at the same time that visual localization is measured. When these joint measure- ments are made in a normally illuminated and structured visual field, all visual localization is as accurate as for the normal, noncurariz- ed individual although EEPI is in error by the large amounts noted above.

Conclusion 1: Since EEPI is differently in error at different eye positions but visual localization is accurate at all these eye positions, it follows that EEPI does not normally influence visual localization in a normally illuminated and structured visual field. The mechanism specified by the Cancellation Theories is thus not normally involved in visual localization for the broad range of conditions of major interest (normal illumination and structured visual fields).

This brief statement is provided early here since the main point does stand out and requires no complicated detail in order to point to the result as a major failure of Cancellation Theory; it will be amplified below. Similarly simple arguments cannot be pre- sented for Type B failures although the arguments for Type B fail- ures have similar force.

The failure of Cancellation Theory for visual localization in normally illuminated and structured visual fields does not imply the nonexistence of "cancellation" altogether (far from it). Thus, although visual localization by the curarized subjects was accurate when viewing was carried out in a normally illuminated and structured visual field, visual localization in darkness was in error by the amount of the error in EEPI (Type A error of visual localization) (Matin et al., 1980, 1981b). From this we infer:

Conclusion 2: Cancellation mediates visual localization in darkness.

Cancellation also plays an important role in (a) intersensory localization in darkness, and (b) sensory/motor localization in darkness. In addition, cancellation may also play some role in intersensory localization in normally illuminated and structured visual fields, and it is also likely that it can contribute to visual localization in illuminated but highly unusually structured visual fields (Sect. 6 below). However, the latter has not yet been demonstrated; experimentation on the matter is in progress. The critical issue with regard to EEPI's influence is whether or not Type A suppression (for which special cases have been previously called "visual capture") is present or not. Just as visual stimulation by a sound source may determine our localization of the sound, overriding auditory cues (e.g. the voice in the movie theater appears to emanate from the screen but may emanate from a speaker 50° away from the screen), so may the visual stimulation of the relation of one object to others influence where that object is seen to be, overriding cues from EEPI.

1.5. Defining visual localization

Before proceeding to specific cases it will be useful to provide a heuristic definition of visual localization. The motive for providing this definition is the desirability of separating a subset of categories labeled visual localization from other subsets of categories of localization involving the visual system that are labeled intersensory localization and sensory/motor localization. Although many commonalities exist, mechanisms subserving each of the subsets also differ from each other in important ways. The basis for the separation suggested here is essentially similar to the suggestion that the field of sensation and perception can be separated into subsets such as vision, audition, somesthesis, etc. Some separate study is desirable although commonalities exist. Although the lines of cleavage between visual localization, intersensory localization, and sensory/motor localization are much fuzzier on closer inspection and possibly not the best ones for use ultimately, providing them here should help to clarify some issues. We first distinguish four classes of *visual localization*; this classification appears to include all important cases of visual localization:

1. A seen object may be visually localized relative to another visual object simultaneously present. Such localizations are given

by reports as "A is to the left of B" or "A is above B". Either A
or B (or both) may be an external object, a view of the observer's
own body, or part of the observer's own body.

2. A seen object may be visually localized relative to an
"internal norm" such as "A is to the left of my median plane" or "A
is above my eye-level horizontal". For present purposes we need not
be overly concerned about what is or is not acceptable as a "norm"
and accept as sufficient for treatment of a norm that a subject
could make the discrimination of A against the norm in question
in total darkness with only A visible[1].

3. An observer can be said to "absolutely identify" the location of a light if, in darkness he is able to correctly name (e.g.,
"number 7") the light when it is illuminated alone from an array of
lights which are identical except for location. Insofar as the observer's judgments depart from chance, some capacity for absolute
identification is manifested. Such absolute identifications could
be treated under (2) above as discriminations against a norm if
observers carry out such identifications by judging a light to lie
at a specific distance from a visual norm. But absolute identifications may be mediated by other means, and until the question of
their mediation is resolved empirically, absolute identification is
best treated separately from matches to a norm.

4. One object may be visually localized relative to a second
although the two are visible only sequentially and not simultaneously. Such sequential visual localization is the central operation
for observing the presence or absence of visual stability when an
eye movement occurs, with presaccadic and postsaccadic views being
the items whose localizations are related.

Given the above classification, all other spatial localizations
for which a stimulus item is visual may then either be classified as
intersensory localization (e.g., "the sound is to the left of the
light") or as *sensory/motor localizations*[2] (e.g., a subject points
his finger or his eye at a light or at a visual norm). (*Intermotor
localizations* such as pointing an eye at a finger in the absence of
visual stimuli are also of interest but will not be discussed here.)
Accuracy of intersensory localization is measured by the accuracy
of the match between stimuli to two modalities (or a norm in one
modality and a stimulus to a second modality); accuracy in sensory/
motor localization is measured by the accuracy of the match between
sensory stimulus (or norm) and motoric positioning (or norm).
Sensory/motor and intersensory localizations also can be categorized
along lines that are both conceptually useful and separate major
differences in experimental paradigm, although we shall not pursue
this here.

2. THEORIES OF PERCEPTUAL STABILITY

Three main classes of theory of perceptual stability may be
distinguished in the literature: This section will introduce them
and briefly relate them to some features of the experimental data.

2.1. Cancellation theories

The common aspect of members of this class of theories is an algebraic and general model of the process. As noted above, these theories assume that perceptual stability is a consequence of the coordination of EEPI and RI. Three variations of this model which suggest different sources for EEPI are schematized in Fig. 1.

Although the main focus of concern regarding cancellation theories has generally been on the source of the EEPI, the concern in this chapter with Cancellation Theories will be with delineating the conditions in which cancellation plays some role from those in which it does not and on some of the characteristics of localization mediated by cancellation. The problem regarding the source of EEPI will be dealt with immediately below in the present section only.

The earliest proposal regarding the source of EEPI is that of Helmholtz (1866) who suggested that an observer visually localized objects by combining information regarding the "effort of will" involved in turning his eyes with information regarding the location of the objects' images on the retina (outflow theory). Other versions of outflow theory were proposed subsequently (von Holst and Mittelstaedt, 1950; von Holst, 1954; Ludvigh, 1952; Merton, 1964). Although many workers have agreed that some second channel of information is required in addition to the pathway leading upstream from the retina, they have not agreed that the "effort of will" constitutes this second channel. Alternative proposals for the second channel make use of peripheral sense organs in the orbit which sense eye position either alone (inflow theory; Sherrington, 1898, 1918; James, 1890) or modulated or combined with signals derived from commands to turn the eye ("hybrid model"; Matin, 1972, 1976a; Shebilske, 1976, 1977)[3]. The term "extraretinal signal" has been employed (Matin, Matin, and Pearce, 1969) in order to refer to such a second channel in a way that remains neutral as to its source and is being employed here as "extraretinal eye position information"(EEPI). This avoids the theoretical commitment generally carried by terms such as "proprioception", "corollary discharge", or, "efference copy" which have been employed for related purposes.

Helmholtz's (1866) basis for outflow as a source of EEPI consisted of three points: (1) He stated that individuals with clinical cases of partially paralyzed or weakened eye muscles reported movement of the visual field when they tried to move their eyes; (2) A push with an external object (e.g. finger) that displaces the eye produces an appearance of movement of the visual field; (3) Attempts to turn the eye in the dark produce visual movements of afterimages but displacement of the eye with an external object does not. James (1890) provided an explanation of the first two observations of Helmholtz by assuming afferent signals regarding eye position from the nonparalyzed eye and dismissed "feelings of innervation" in general (including Helmholtz's "effort of will").

Although Kornmuller's (1930) report of experiments with partial paralysis supported the observations regarding perception with

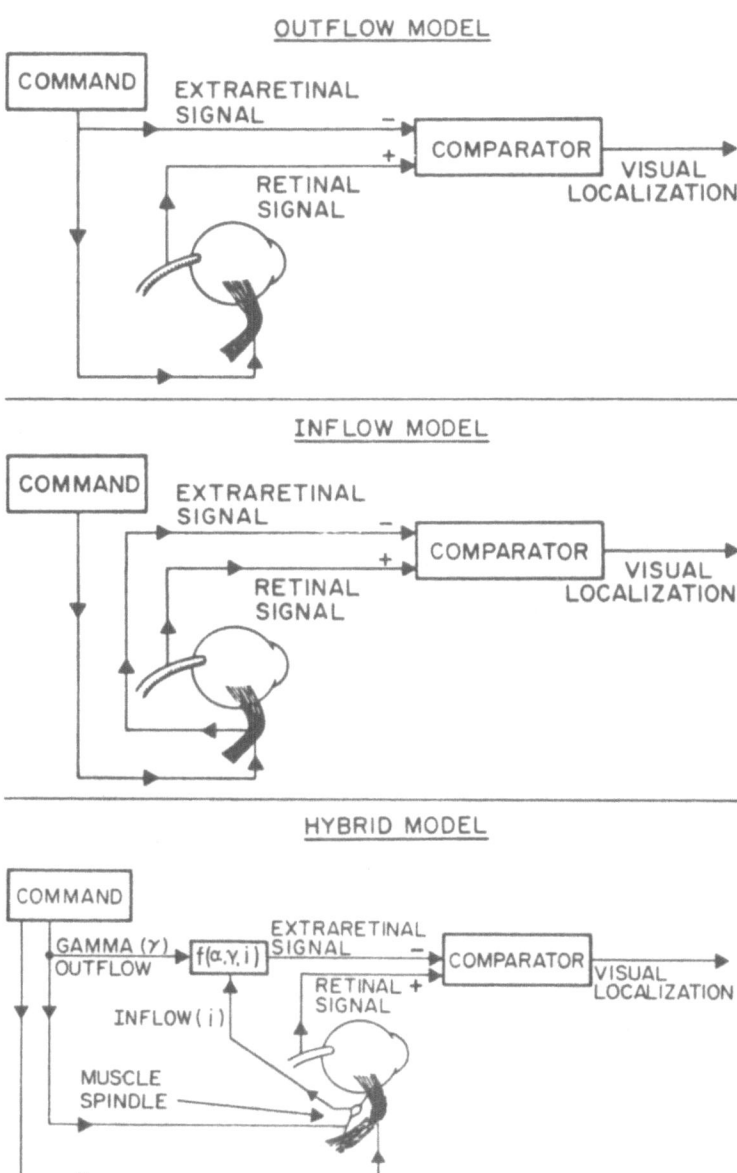

Fig. 1. Three varieties of Cancellation Theory developed to explain
        stability of visual localization in the presence of eye
        movements are schematized. All three assume that extraretinal
        eye position information (EEPI) and information regarding
        shift of the image at the retina (RI) are added by a neural
        mechanism (comparator) whose output is "visual localization".

clinical paralysis, four more recent reports (Siebeck and Frey,
1953; Siebeck, 1954; Stevens, Emerson, Gerstein, Kallos, Neufeld,
Nichols, and Rosenquist, 1976; Brindley, Goodwin, Kulikowski and
Leighton, 1976) have agreed that under total experimental paralysis
of the extraocular muscles no appearance of movement of the visual
field is seen as is required by outflow theory. The paralyzed eye
observations reported by Helmholtz and later by Kornmuller were
most often cited as critical support for outflow theory. This repeated failure to observe movement would seem to throw support in
the direction of either inflow or hybrid theory (see Matin, 1972,
1976a, and Shebilske, 1977 for more detailed reviews of work on the
question). However, just as previous reports of movement in a paralyzed eye are not critical support of outflow theory (Matin, 1972,
1976a; Shebilske, 1977), so is "no movement" not critical for deciding on the source of EEPI. Experimental support for both inflow and
outflow involvement in visual localization does exist. The discussion below, particularly with regard to Type A failure of Cancellation Theory, also provides a basis for suggesting that the question
regarding the source of the EEPI as it might be involved in visual
localization should be considered entirely open until experiments
with total paralysis are carried out in total darkness; such experiments are in progress at this time.

2.2. Information-theoretic or cognitive approaches

Several workers have described approaches regarding perceptual
stability that have either a cognitive or information-theoretic
flavor. MacKay's (1962, 1972, 1973) is one of these and I shall
comment on it here: He rejects a physiologically oriented cancellation mechanism, suggesting instead that the individual "evaluates"
relevant information. Thus: "Although the sensory mechanism cannot
properly interpret sensory changes without some information as to
the activity of the motor system, from an information-engineering
standpoint the need is not for the changes due to voluntary movement
to be *eliminated* from the sensory input, but for them to be appropriately *evaluated* by the central mechanism responsible for the
organism's "conditional readiness" to reckon with its environment"
(MacKay, 1973, p. 308).

In today's psychological terminoloy "evaluation" would generally
be employed to specify "cognitive" processing. MacKay's use is as
follows: " 'Evaluation' here does not entail (or deny) conscious
mental activity, but refers simply to the kind of computing process,
familiar in feedback-guided automata, in which sensory signals are
compared against criteria, so as to generate feedback signals indicating whether the states of affairs signalled are 'satisfactory' or
otherwise, and what corrective or reinforcing actions are required,
if any (MacKay, 1973, p. 316)".

The approach has the difficulty common to information-theoretic
models: it offers a format but not a testable theory. What is required for its proof or disproof is unclear. For example, although MacKay

opposes "evaluation" to "subtractive cancellation", his definition
of evaluation does not appear to preclude subtractive cancellation
being a form of evaluation or a portion of an evaluation process.
For MacKay's statement to become a useful theory it is necessary
that "evaluation" be defined so that it becomes possible to tell
what is a case of evaluation and what is not. As it has been presented it appears to cover any and every operation an observer may carry
out on sensory input, and we are not differently informed by the
assertion that the observer "evaluates" than by saying that he "perceives".

Elsewhere MacKay (1962, p. 101) also states "... that we should
regard perception of change as the achievement requiring information
to justify it, with stability as the null hypothesis," that what the
observer perceives quite generally depends on what the observer's
goals are, and that the relation between the observer's goals and
perceptual stability is the outcome of an active process mediated by
"hypotheses" which the observer constructs and tests. This does
provide a theoretical assertion that is a relative of the one made
by Helmholtz's outflow theory, but differs from it in suggesting
that success or failure in reaching a goal is as important in determining perceptual stability as is the attempt to do so. Although
evidence supporting an involvement in the localization-mediating
process of an observer's goals has been demonstrated, (Pola, 1972,
1976, Pola and Matin in prep., see Sect. 5 below) it is also clear
that some important aspects of the process controlling perceptual
stability have no more to do with the goals of an individual's eye
movements than does the fact that coal normally appears equally
black in both sunlight and dim illumination although the light
entering the observer's eyes from the coal is orders of magnitude
different in the two cases. Although it would be possible to pursue
this point regarding the primitiveness of mechanisms for preserving
perceptual constancy at length here (and a number of aspects of the
data considered below are quite conclusive), it suffices to note:
(1) 10 minutes (and probably less) after birth a human infant's eye
movements are correctly directed toward the source of a sound
(Wertheimer, 1961). (2) An individual who from birth sees no form
but only a uniform haze in the brightest sunlight reports a phosphene
to his left when the right side of his eye is pressed and vice versa
(Schlodtmann, 1902); this localization fits our expectations from
knowledge of the neuroanatomy of the visual system and our more
common beliefs about the connection of neuroanatomy to function, but
it does not fit notions about "hypotheses", whether or not they are
assumed to have any flexibility to them.

2.3. Dual-suppression theory of spatial localization

An approach to the perceptual stability problem has been developed in a number of reports (c.f., Matin and Pearce, 1965; Matin
et al., 1969; Matin, 1972, 1976b, 1981; Matin et al., 1980, 1981b)
to deal with the results of experiments with visual localization in

the presence of voluntary saccades and during steady fixation which
now has sufficient coherence to be given a name (Dual-Suppression
Theory of Spatial Localization). It appears to account for the main
results discussed in this chapter, incorporates a cancellation
mechanism but is indifferent regarding the source of EEPI, and yields
some predictions described in Section 6 below. It may be summarized
as follows: (1) Spatiotemporal relations among visual stimuli are
important determiners of perceptual stability; (2) EEPI is of limited
utility in the production of perceptual stability (more limited than
proponents of Cancellation Theory have recognized). The limitation
is threefold: (a) cancellation does not normally influence visual
localization in normally illuminated and structured visual fields;
use of EEPI for visual localization is suppressed under these conditions (Type A suppression); (b) where cancellation does apply visual
localization is much less accurate, and (c) much less precise
(several orders of magnitude) than visual localization mediated by
spatiotemporal relations within the visual field. (3) The "shift"
of coordinates produced by EEPI in the interest of maintaining perceptual stability normally brings the "present view" of a visual
field into correspondance with the memory of a previous view at a
previously held gaze direction. (Although I had previously (Matin
et al., 1969) considered it possible (though less likely) that EEPI
operated on the memory of the previous view rather than the present
view, the occurrence of Type A a/v error with partially paralyzed
observers requires that EEPI operate on the present view.) (4) Saccadic suppression (Type B suppression) is mainly a process in which
the spatiotemporal properties of the retinal image of any visual
target interferes with and eliminates from perception (a) perception
of the "smeared" image of itself that is produced by the saccade,
and (b) perception of the persisting view of itself preceding the
saccade. (5) Normal visual persistence of previous views is shortened
by saccadic suppression allowing the view available immediately preceding and during a saccade to be eliminated from the "visual
present", what was perceived recedes into a long term memory bank
where it does not interfere with the postsaccadic view; the difference between this memory and the present view can be operated on
by EEPI. (6) Cancellation mediates visual localization in darkness;
it also mediates sensory/motor, intersensory, and intermotor coordination under conditions whose breadth and range have yet to be
specified although these conditions can include both illumination
and darkness.

The limitation on EEPI utilization has been found to be even
greater than originally anticipated. Earlier approaches to dealing
with mechanisms underlying perceptual stability have dealt with the
influence of a structured visual field on visual localization, as if
this influence could be added to the picture after the decision was
made regarding which version of the cancellation mechanism (Fig. 1)
was true. The recent work with partially paralyzed subjects (Matin
et al., 1980, 1981b) demonstrates however, that the original experiments that were designed to determine the source of EEPI will have

to be redone under conditions in which care is taken to control for
the presence or absence of illumination; such experiments are in
progress.

3. LOCALIZATION WITH STEADY GAZE

3.1. Normal observers

3.1.1. Visual Localization Relative to Visual Norms
     With steady gaze at a fixation target employed to fix eccentri-
city of the position of the eye in the orbit (gaze direction) normal
subjects reliably set a second light to perceived eye-level horizontal or to the median plane. Although systematic departures of
average settings from accuracy are observed, the average setting to
the perceived position of eye-level horizontal changes little with
variation in vertical gaze direction; average setting to the median
plane changes little with variation in horizontal gaze direction.
These settings made in darkness (Fig. 2; reproduced from Matin et
al., 1980, 1981b) are themselves a demonstration of the involvement
in visual localization in dark fields of an extraretinal channel
providing information regarding the position of the eye in the orbit
(EEPI). If such information were not available, the orderly change
in retinal location of the visual stimuli set to the norms would
not be produced with variation in gaze direction, and the norm settings for normal subjects in Fig. 2 would not manifest the stability
they do.
     The reliability of the settings to the above norms is about ±2°,
values that provide some indication of the order of reliability of
the EEPI that is involved in visual localization. The variable
errors in repeated settings are not due to coarseness of the retinal
information with increasing retinal eccentricity of the light set to
the norm. This is shown by the fact that variability in setting is
not noticeably increased by increasing eccentricity of gaze although
such an increase does increase the retinal eccentricity. A similar
statement does not apply to the systematic errors of accuracy of the
average settings; these errors may in part be related to distortions
between retinal distances and perceived distances and in part to
distortions between EEPI and actual eye position. Considerable evidence for the presence of distortions between retinal and perceptual
distances is available (c.f., Helmholtz, 1866, Ogle, 1950; Pearce
and Matin, 1969). Further work will be required to separate the influence of retinal from extraretinal distortions on setting accuracy
and to measure the extent of distortions in EEPI.
     In order for settings of the visual target to be made to eye-
level horizontal (Figs. 2b and 2e) it is not only necessary that
EEPI regarding angle α (Fig. 2) be combined with information regarding retinal image location of the visual target, but with information regarding orientation of the head relative to gravity (angle β,
Fig. 2) as well. Different values of β do not systematically in-

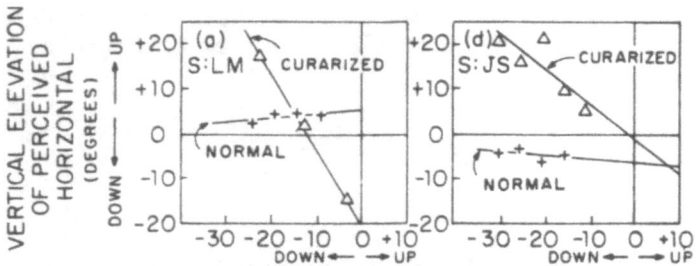

VERTICAL ELEVATION OF PERCEIVED HORIZONTAL (DEGREES)

VERTICAL FIXATION DIRECTION: ANGLE α (RELATIVE TO HEAD)
(DEGREES)

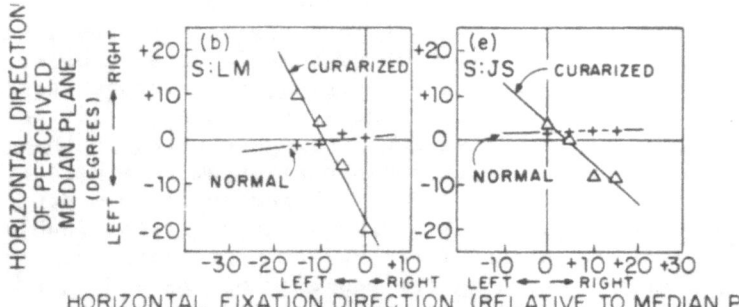

HORIZONTAL DIRECTION OF PERCEIVED MEDIAN PLANE (DEGREES)

HORIZONTAL FIXATION DIRECTION (RELATIVE TO MEDIAN PLANE)
(DEGREES)

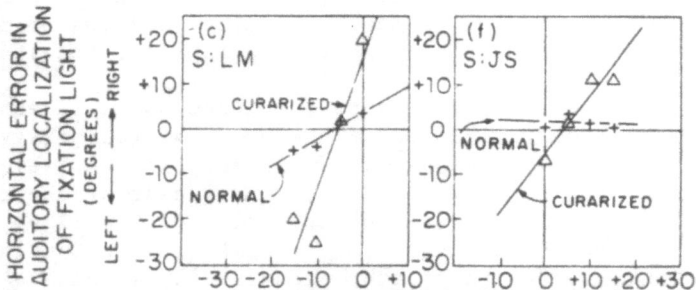

HORIZONTAL ERROR IN AUDITORY LOCALIZATION OF FIXATION LIGHT (DEGREES)

HORIZONTAL FIXATION DIRECTION (RELATIVE TO MEDIAN PLANE)
(DEGREES)

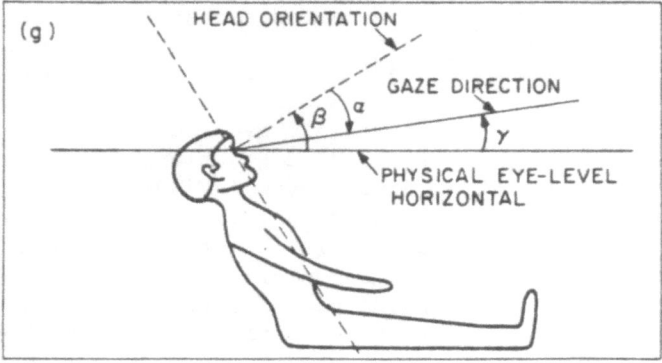

Fig. 2. (From Matin, Picoult, Stevens, Edwards, Young and
MacArthur, 1980, 1981b.)
In all figures + represents measurements made on the normal
unparalyzed observer. Δ represents measurements made on the
observer in a paralyzed condition. The straight lines were
fitted to each set of data employing a least squares
criterion.
(a) and (d). *Perceived Eye-level Horizontal:* In complete
darkness the observer (LM or JS) fixated a small visible
target whose angular elevation with respect to the transverse plane through the head ("head orientation") is plotted
on the abscissa; this elevation defined the vertical angle
of the eye in the head (angle α). The transverse plane
through the head was itself above the physical horizontal
by 24° for LM and 30° for JS (angle β). While maintaining
fixation on this first light the observer instructed the
experimenter to set a second peripherally-viewed light
(which was movable in the same vertical meridian as the·
first target) to a height that appeared to be at his eye-
level. The height of this setting is plotted on the ordinate as vertical elevation of perceived horizontal.
(b) and (e). *Perceived Median Plane:* The observer (LM or
JS) fixated a small visible target in complete darkness
whose horizontal angular deviation from the physical median
plane through his body is plotted on the abscissa. This
angle defined the horizontal angle of gaze. While maintaining fixation on this first light, the observer instructed
the experimenter to set a second visible target (which was
movable in the same horizontal plane as the first target)
to the perceived median plane. The horizontal eccentricity
of this setting is plotted as the ordinate.
(c) and (f). *Auditory/Visual Matches:* The abscissa is the
horizontal location of the fixated light with respect to
the median plane of the observer. The ordinate is the error
in the auditory localization of the fixated light (the
difference between the location of the fixated light and
the physical location of the sound). For both Figs. 2c and
2f normal room illumination was present. The results are
indistinguishable when the experiment is carried out in
total darkness (see Fig. 3).

fluence the settings to perceived eye-level horizontal, nor are the
results in Fig. 2b and 2e modified when the fixation target is the
only visible target and is itself set to the perceived eye-level
horizontal (Matin et al., 1980, 1981b).

3.1.2. Visual Localization of two simultaneously-presented targets

When two visual targets are presented simultaneously sensitivity
for their relative localization is extremely good. Under conditions
designed to maximize sensitivity, thresholds for detection of vernier offset as small as 2" of visual angle are obtained (an impressive manifestation of the complexity of neural processing of the
visual system behind a photoreceptor layer whose grain size is about
25"). Since eye movements do not influence the relative retinal
locations of two simultaneously-viewed targets any cancellation
mechanism involving EEPI would not be expected to produce systematic
distortions. We would not expect considerations regarding EEPI to
be significant here. These expectations are borne out insofar as
acuity for vernier offset is at most only slightly different under
image stabilization as compared to normal viewing (Keesey, 1960;
Fender and Nye, 1962). Further, although Rattle and Foley-Fisher
(1968) report longer intersaccadic intervals are related to better
vernier acuity (a result that is likely to be due to the increased
duration available for unimpeded temporal integration) Krauskopf,
Graf, and Gaarder (1966) report no change in vernier acuity for
brief flashes measured in close temporal proximity to an involuntary
saccade.

3.1.3. Visual localization of two sequentially-presented targets

However, when two targets are presented successively the eye
movements in the interval between their presentations contribute to
the offset between their images at the retina. Considerations regarding the possible involvement of EEPI then become of substantial
concern although for steady fixation no influence of EEPI is observed:

When subjects report whether the second of two foveally-present-
ed flashed targets (small discs, vertically-oriented lines of a
vernier target) lies to the left or to the right of the first of
the pair, the variable error (1/acuity) increases monotonically with
the time interval between presentations of the two targets, reaching values between 5' and 25' at 1 to 3 second intervals and showing
further increases to 40' at intervals as long as 32 seconds (Matin,
Pearce, Matin, and Kibler, 1966; Matin and Kibler, 1966; Matin,
1972; Matin, Pola, Matin, and Picoult, 1981a; Fiorentini and Ercoles,
1968; Foley, 1976, 1978; Findlay, 1974). These errors in visual
localization bear a very marked systematic relation to the involuntary eye movements which occur in the dark interval between presentations of the two targets, a relation that is largely in correspondence with the direction and magnitude of offset of the two targets
at the retina. Thus, for example, even when the second flashed vertical vernier line is substantially to the left of the first line,

if the eye has moved sufficiently far to the left in the dark interval, the subject reports the second line to lie to the right of the
first. As much as 2/3 to 3/4 of the variance of the psychophysical
reports can be predicted from the relative retinal offset of the two
targets calculated from the conjunction of measurements of eye movements and the values of physical offset between the two targets
(Matin et al., 1966, 1981a).

Three other conclusions from these experiments are relevant
here:

(a) EEPI related to the involuntary changes in eye position
during the dark interval does not influence the sequential localization discrimination systematically. Although these changes in eye
position may be as large as 3°-4° or more in a three-second dark
interval and influence the retinal offset between the images of the
two targets as noted above, the perceptual mechanism responsible
for the localization discrimination does not compensate for the eye-
movement produced retinal offset. This result is compatible with
either the outflow or the hybrid model, but not the inflow model
(Fig. 1). Interestingly, an eye-movement-produced change in retinal
image location in as short an interval as 2 msec is sufficient to
produce a change in discrimination performance of a vernier task
with lines flashed sequentially (Matin et al., 1981a).

(b) There remains a substantial error variance that is unac-
counted for by the change in retinal image location due to the un-
compensated eye movements. This unaccounted-for error variance increases with the interstimulus interval and thus represents either
(1) noise in EEPI (i.e., fluctuations in the correspondence between
actual eye position and eye position as signaled to the mechanism
determining visual localization), or (2) local sign noise in the
visual system unrelated to EEPI but related to a coarsening and
distortion of the map relating perceived space to the retina (Matin
et al., 1966; Matin et al., 1981a). Either would be a form of
memory loss. (Also see Kinchla and Smyzer, 1967; Kinchla and Allan,
1969; Kinchla, 1976.)

(c) Whatever this variable error in the sequential localization
task is due to, it is clearly smaller (by more than an order of
magnitude) than the extraretinal error related to eye position that
is involved in maintaining fixation in darkness (Cornsweet, 1956;
Matin et al., 1966, 1981a; Matin, Matin, and Pearce, 1970;
Skavenski and Steinman, 1970; Skavenski, 1971; Allik, Rauk, and
Luuk, 1981). In addition: (1) fixation stability in darkness is 100
times better against a visible fixation target than without a fixation target, and (2) localization errors with successive presentation
are as much as 1000 times greater than are obtained with simultaneous presentation. These points are conclusive evidence regarding
the relative insensitivity of EEPI in comparison to the sensitivity
of information derived from the retina.

3.1.4. Intersensory localization

When visual and auditory localizations are made simultaneously

under normal conditions discrepancies between them are not normally
noticed, but instead are eliminated by the visual capture of auditory localization. It is only necessary to recall that the source of
the sound which we localize at the speaker's lips on the movie screen
may be located as much as 90° or more away from the speaker's lips
on the screen to recognize the power of visual capture. Some of the
bases for visual capture and other visual/auditory interactions have
been delineated (although not without controversies) (c.f., Witkin,
Wapner, and Leventhal, 1952; Jackson, 1953; Platt and Warren, 1972;
Jack and Thurlow, 1973; Thurlow and Jack, 1973; Choe, Welch, Gilford
and Juola, 1975; Jones and Kabanoff, 1975; Bertelson and Radeau,
1976; Shelton and Searle, 1980). Temporal synchrony between visual
and auditory stimuli promotes visual capture, and although there is
some disagreement about whether the presence of illumination improves auditory localization, there is some agreement that freedom
to move the eyes during the localization task can improve acuity for
auditory localization in illumination and darkness. Under conditions
in which visual capture can play no role, localization matches between light and sound are generally fairly accurate for most individuals, but may manifest substantial constant errors. Although not
all of the references referred to here agree, we have found these
errors to be identical whether carried out in darkness or in normal
illumination (Picoult, MacArthur, Young, Stevens, Edwards, and
Matin, 1980). The reliability with which observers report whether
one sound source lies to the left or right of a second sound source
when neither source is seen is about 2° (Mills, 1958, 1960). When
visual/auditory matches are made under darkroom conditions in which
subjects match the location of a sound to the location of a single
light by choosing one of an array of randomly presented speakers
arranged around the subject, reliability is also about 2° (Picoult
et al., 1980).

Individuals differ, however, on the accuracy of the auditory/
visual match. Figure 3 shows three characteristic forms for the
matches as a function of gaze eccentricity of the fixation target:
(1) Some individuals (DY, RM, Fig. 3) make essentially veridical
matches throughout. (2) For some individuals (EP) the location of
the sound that is set to match the light is roughly a fixed angular
distance from the light at all gaze eccentricities (Type C a/v
error); (3) For some individuals (LM) the error in the match between
sound and light changes regularly with gaze eccentricity, yielding
a slope other than 1 in a plot such as Fig. 3 (Type A a/v error).

Although it is not logically necessary, in almost all cases
individuals with Type C a/v error have been found to possess binau-
ral asymmetry in auditory sensitivity, and adult individuals selected
for such asymmetry have also manifested Type C a/v errors regardless
of whether the asymmetry was acquired in adulthood or had been present since childhood, a result which strongly suggests that the relation between auditory and visual maps of space is innate and essen-
tiallyunmodifiable when visual capture is eliminated (Picoult et al.,
1980). The existence of an innated correspondance between visual and

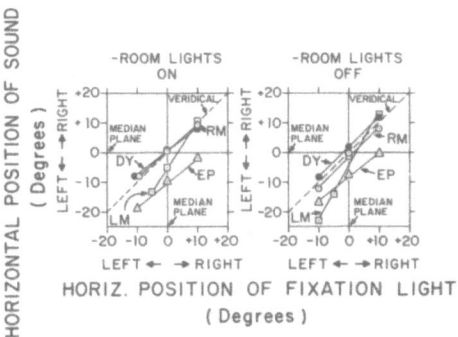

Fig. 3. (From Picoult, MacArthur, Young, Stevens, Edwards and Matin,
        1980.)
        Auditory localization of a fixated light for various hori-
        zontal positions of the light. The ordinate is the PSE set-
        ting of the sound to the perceived horizontal location of
        the light. Data is for the same four subjects in normal room
        illumination (left panel) and in darkness (right panel).

auditory maps of space is further supported by the fact that 10
minutes after birth infants turn their eyes in the correct direction
to a sound source (Wertheimer, 1961). Normal adult individuals whose
hearing has been attenuated by an earplug in one ear also show Type
C a/v errors in the direction predicted by assuming that the monaural
reduction in effective stimulus intensity produces an increase in
neural response latency that acts to produce a time-of-arrival difference at the neural center processing auditory localization, a
time difference that is equivalent to one that can be produced
naturally by moving the sound source in the horizontal plane
(Picoult et al., 1980; cf., Green, 1976 for discussion of intensity-
time tradeoff in auditory localization). Although Type C a/v errors
should in principle (and may yet) be observed in individuals with
phorias and strabismus, examination for this has not yet been carried
out.

Type A errors appear to be a consequence of a scale distortion between the magnitude of eye deviation from some primary position of gaze and the EEPI that is involved in visual localization (Picoult et al., 1980; Matin et al., 1981b). This inference is strongly supported by the experiments with experimentally paralyzed individuals discussed below.

The stable and apparently normally unmodifiable relation between visual and auditory maps of space in all individuals (whether veridical, or manifesting Type A or Type C errors) is surprising in view of the great flexibility manifested by auditory localization under normal conditions which promote visual capture. Since the relation of ocular gaze direction to norms of visual localization is modifiable (Kalil and Freedman, 1966; McLaughlin and Webster, 1967), the stability of the relation between auditory and visual maps must be a consequence of the joint operation of two factors: (a) Any experience-mediated modification between ocular gaze direction and norms of visual localization is *toward* bringing visual norms into correspondance with particular planes of the body (e.g. correction of the relation of ocular gaze direction to the perceived median plane when a displacing prism is worn is toward bringing the physical direction that is visually perceived as the median plane toward the body's true median plane); (b) The map of auditory location to physical location is not normally modifiable by experience (also see Hay and Pick, 1966).

3.1.5. Sensory/motor localization

An observer in darkness may look (motor) at a single light (visual sensory), look (motor) in the direction of his own median plane (visual norm), look (motor) at a sound (auditory sensory), look (motor) at his finger (proprioceptively sensed finger position), or point his finger (motor) at a light (visual sensory). Obviously other relations of sensory and motor localizations may be noted. We will only comment on the first two cases.

a. *Looking at a Single Light or Attempting to Maintain a Previous Fixation Position.*

Variability of the position of the eye when an observer repeatedly turns his eye to look at a given stationary target or gazes at it continuously is extremely small. Standard deviations of eye position during steady fixation are 1'-3' of arc visual angle (see Ditchburn, 1973, for review and summary). This corresponds to 3 to 9 cone diameters. This extremely fine ability clearly depends on processing of information that is derived from the position of the visual image on the retina and not on EEPI, a point first made most clearly by Cornsweet (1956), who showed that removing the fixation target while requiring the observer to continue holding the same eye position yields an increase in positional variability of 2 to 3 orders of magnitude and substantial systematic errors which change with time as well. This has been supported by subsequent work (Nachmias, 1961; Fiorentini and Ercoles, 1966; Matin, Matin and Pearce, 1970; Allik, Rauk and Luuk, 1981; Skavenski, 1971; Skavenski

and Steinman, 1970) who report increases in fixation errors at rates
that range between 2'/sec and 10'/sec. In darkness the eye may reach
distances as far as 10° in any direction from the desired position
with a standard deviation of 2° to 5° during attempts at maintaining
a fixed position extending over a minute of time (Figures 6-8 of
Allik et al., 1981 are particularly instructive here); departures
from a random walk are measureable early in the dark interval and
more substantial later. The basis for the fact that we "look" with
our central fovea has often been considered to be the fact that
acuity peaks in the central fovea. An alternative, that the central
foveal cones are "preprogrammed for looking" regardless of acuity
has never been investigated. It is not yet clear whether variability
in eye positioning in relation to a visual target is due to variability in motor control or to variability in the retinal point with
which the subject attempts to "look" although the evidence is
strongly in favor of the former. The fact that acuity for vernier
offset normally can reach values as small as 2"-10" indicates the
fineness of the resolution with which the visual system itself is
capable of processing spatial localization. This fineness in visual
acuity supports the conclusion that the variability in attempting
to look at a fixed point in darkness is in the motor control of the
eye and not in the visuosensory system.

   b. *Looking in the Direction of Visual Norm.*
   Variability in the direction with which an observer points his
eye when instructed to look in the direction of his median plane
in darkness is normally about 2°-5°. This direction is capable of
considerable modification, however, as is demonstrated by its susceptibility to adaptation by wearing of wedge prisms which laterally
displace the visual field (Kalil and Freedman, 1966, McLaughlin and
Webster, 1967; for reviews dealing with questions related to such
adaptation see Kornheiser, 1976; Welch, 1978). Skavenski (1972) has
reported that observers in darkness looking in the direction of their
median planes when their eyes are loaded by an external force pulling
on a stalk attached to a contact lens can discriminate the direction
of pull when its magnitude is 7°-14°. They can also return their gaze
to the median plane with an accuracy of 2°-3°. These results demonstrate the control by inflow information of eye position during
steady gaze in a given direction. Although it is clear that outflow
EEPI makes some contribution to eye position control from the fact
that we can voluntarily turn our eyes to the left or right in darkness, what further contribution outflow EEPI makes has yet to be
ascertained. Brindley and Merton (1960) indicated that their subjects could not even report eye turns as large as 40° produced by
the experimenter's pulls on the subject's extraocular muscles; this
suggested no inflow involvement. This result is considerably discrepant from Skavenski's. However, Brindley and Merton's subjects were
not attempting to maintain fixation in a given direction. It has
been suggested (Matin, 1972) that Skavenski's (1972) and Brindley
and Merton's (1960) apparently inconsistent results are compatible
if we assume that gamma bias to the muscle spindles is set by the

attempts of Skavenski's subjects to look in a given direction while
the lack of a similar attempt in Brindley and Merton's case removed
a useful input from their observer's spindles.

3.2. Experimentally Paralyzed Observers

3.2.1. Visual Localization Relative to Visual Norms

When the extraocular muscles are weakened by systemic injection
of curare the extent of gaze in every direction is reduced (Matin
et al., 1980, 1981b). This limitation on gaze eccentricity is increased by increased levels of curarization. However, within the
range for which variation in ocular gaze direction is still possible,
no abnormalities of visual spatial localization are noted by subjects while gazing steadily at any particular point in a normally
illuminated and structured visual field; appearances are no different
than without partial paralysis. This normality of appearance holds
whether or not parts of their own body are seen. However, when all
illumination is extinguished leaving visible only a fixation target
and a test target, very substantial errors in localization of these
targets occur with a magnitude that increases with level of paralysis (Matin et al., 1980, 1981b). At a given level of paralysis,
errors in setting the test target to the perceived location of the
median plane increase linearly with horizontal deviation of gaze
direction from some "zero" position of the eye in the head (Fig.
2b, e). Errors in setting the test target to the eye-level horizontal
increase linearly with vertical deviation of gaze direction from
some zero position of the eye in the head (Fig. 2a, d). These errors
are unchanged by changes of head-and-body tilt around a horizontal
axis through the two ears[4], indicating that the errors are not a
consequence of distortions of information about the direction of
gravity[5]. Thus both errors in setting the target to the median
plane (settings made in a plane for which there is no polarization
relative to gravity) and to eye-level horizontal are a consequence
of the increased demand required to produce a given eyeturn by means
of a system in which the efficiency of neuromuscular transmission
has been reduced (the curare reduces the magnitude of muscular con-
traction produced by a given neural bombardment at the neuromyal
junction). The errors in localization are a consequence of errors
in EEPI, errors which appear to treat the eye as if it was in a
position to which the level of effort exerted normally would have
turned it without curarization. These results cannot be accounted
for by the inflow model (Fig. 1b) but are effectively dealt with
either by the outflow (Fig. 1a) or hybrid models (Fig. 1c). This
conclusion leaves the question regarding the source of the EEPI in
the same uncertain condition as indicated earlier in this review
and in previous reviews (Matin, 1972, 1976a).

*However, the errors in Figs. 2a, b, d, and e are only present
in darkness. In normal illumination visual localization by the
experimentally-paralyzed individuals is entirely accurate (Table I).*
Thus, for example, although in darkness EEPI may be in error by +30°

Table I. Dependence of localization errors of oculoparalytic illusion (OPI) on presence or absence of visual field for the partially paralyzed observer.

Condition of Room Illumination	Visually Perceived Median Plane Setting	Visually Perceived Eye-level Horizontal Setting	Auditory/Visual Match of Localizations	Naming Auditory Directions	Pointing to Proprioceptive Horizontal	Pointing to a Visual Target
Total Darkness	Gaze-dependent errors	Gaze-dependent errors	Gaze-dependent errors	Accurate	Accurate	Gaze-dependent errors
Structured Visual Field	Accurate	Accurate	Gaze-dependent error	Accurate	Accurate	Accurate

in one gaze direction and -20° in another gaze direction as indicated
by the errors in visual localization, in normal illumination at these
same gaze directions visual localization is perfectly accurate. The
two conclusions forced by these results are the ones which we noted
above: *(a) The cancellation mechanism of Fig. 1 does not normally
influence visual localization during steady gaze in normal illumination. (b) A cancellation mechanism involving EEPI is the main basis
for visual localization relative to norms in darkness.*

3.2.2. Intersensory Localization

When subjects who are partially paralyzed for experimental
purposes match locations of auditory and visual targets in a horizontal plane, substantial errors are made whose magnitude increases
with level of paralysis (Matin et al., 1981b). The auditory target
is set more eccentrically than the visual target to which it is
matched. These Type A a/v errors vary systematically with horizontal
gaze direction, bearing the same quantitative relation to gaze
direction as do Type A errors of visual localization as manifested
by the median plane settings of visual targets in darkness for the
partially paralyzed observer (Fig. 2). *However, whereas the median
plane settings of visual targets are only in error in darkness,
identical auditory/visual mismatches are made in a normally illuminated visual environment as in darkness.* These auditory/visual
errors only occur, however, under conditions in which visual capture
of the auditory target is eliminated. The auditory system is not affected by the curarization (Matin et al., 1980, 1981b), and the
auditory/visual mismatches are not a consequence of auditory errors.

Thus, although no signs of the errors in EEPI are obtained in
measurements of visual localization in the presence of normal illumination (perception of visual target location in relation to
norms; localization of one visual target relative to a second), it
is clear that the errors in EEPI themselves have not been corrected,
but that EEPI is simply not involved in visual localization when
viewing is in a structured visual field. The specific aspects of
the structured visual field which act to suppress utilization of
EEPI for visual localization have not yet been investigated. The
schematic representation of Fig. 4 will assist in clarifying some
of the relations that have been described.

3.2.3. Sensory/motor Localization

When the observer cannot see his arm it is clear that past-
pointing to visual stimuli does occur in subjects who are partially
paralyzed by curare (Stevens, 1978). How much of this is due to
errors in visual localization and how much to errors in the mechanisms controlling the limbs is not yet clear; certainly a large
portion of the effect must be a consequence of visual errors. However, a partially paralyzed subject who can barely raise his fore-
finger can point it accurately along the proprioceptively perceived
horizontal and can do so regardless of the orientation of his body
relative to gravity (Matin et al., 1980, 1981: Table I). Although

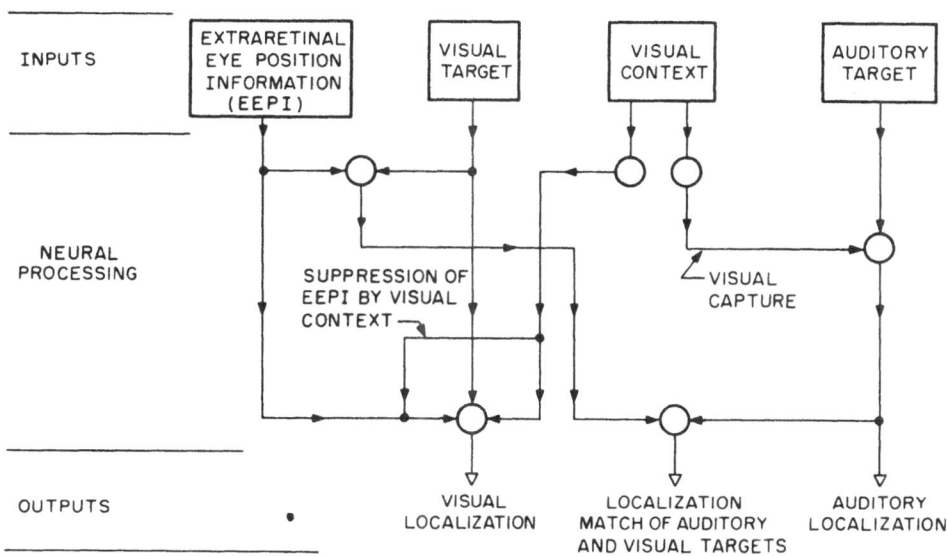

Fig. 4. Roles of EEPI and visual context in determining visual loca-
        lization and in matching auditory and visual localizations.
        Visual context suppresses utilization of EEPI for determin-
        ing visual localization while simultaneously allowing it to
        be used for matching auditory and visual localizations.
        Visual capture of auditory localization is determined by
        different aspects of visual context than are involved in
        suppression of EEPI utilization in visual localization;
        hence visual capture can be eliminated separately from
        visual-context-determined suppression of EEPI (as in the
        present experiments). Direction of information flow is
        shown by the solid arrows; open circles represent processing
        of neural information.

a considerable controversy continues regarding the relative contributions of inflowing and outflowing information to control of the
limbs (cf., Evarts, 1971; Goodwin, McCloskey, and Matthews, 1972;
Roland, 1978; Shebilske, 1978), the latter observation leaves no
doubt about control by inflowing information. Since the effort to
raise the finger is near the upper limit of the effort that the
subject is able to muster, and so very different from the effort in
raising the finger when he is not paralyzed, inflowing information
must be involved.

The likelihood that visual errors are the main basis for past-pointing in partially paralyzed subjects is increased markedly by the observations of a totally paralyzed observer (Stevens et al., 1976): When he thought of trying to reach for a particular visible object while he was attempting to turn his eyes in one direction, although no changes were visually perceived, the observer felt that he would have to reach in a very different direction than if he were attempting to turn his eyes in a different direction; of course, in this case the eyes remained pointed in exactly the same direction during both attempts (Stevens et al., 1976).

4. ON SEPARATING VISUAL LOCALIZATION FROM SENSORY/MOTOR LOCALIZATION

There are several important reasons for distinguishing experiments on visual localization from those on sensory/motor localization. Most important is the fact that visual localization of an object may diverge considerably from its localization by a motor response directed at the object. This is readily exemplified by the fact that an individual who sets a foveated visual target so that it appears to lie in his median plane will set the target to a very different physical location immediately following removal of a laterally displacing (wedge) prism to which he has adapted than he will after the extinction of adaptation to the prism. Foveation (motor response) to these very different physical locations is thus accompanied by perception of them as lying in the identical visual direction. Such adaptation consequent on change in the relation between ocular orientation and visual localization (Kalil and Freedman, 1966; McLaughlin and Webster, 1967) implies that EEPI utilized to position the eye ("sensory/motor localization") has been processed differently or is itself different from EEPI involved in visual localization.

In spite of the difference in the relation between foveating direction and visual direction consequent on adaptation to prismatically-induced lateral displacement, the accuracy of foveating saccades between visual targets is essentially unchanged in different states of perceptual adaptation. Although this is not at all surprising, it is a further clear indication of the separability of visual localization and sensory/motor localization.

5. LOCALIZATION IN THE PRESENCE OF SACCADIC EYE MOVEMENTS

5.1. Introduction

As for the case of steady fixation discussed above, visual localization refers to discrimination in which the location of a visual target is judged relative either to a visual norm or to a second visual target or in·which the observer absolutely identifies stimulus location. However, here we shall only have experiments to

deal with in which the location of one visual target is judged
relative to a second one, since experiments have not been reported
for saccades involving either absolute identification or judgments
relative to a visual norm.

A number of investigations have probed the influence of EEPI
on visual localization in the presence of saccadic eye movements by
measuring an observer's visual localization of a brief test flash
(TF) whose time of occurrence bears a known temporal relation to the
saccadic eye movement. In some of the experiments the TF was elec-
tronically triggered by an electrical signal derived from apparatus
monitoring the eye movement; in other experiments the TF and
saccade occurred independently and their temporal relation was
derived from recordings of eye position. Since locations of the
eye and the visual target producing the flash were separately
variable in the latter experiments, calculations by the experimenter
were usually required to determine the retinal location at which the
TF was imaged. In all of these experiments, visual localization of
the TF was reported by the subject in relation to some other visual
target that was available for some known time either before, simultaneously with, or after the TF was presented.

It is worth noting here that some important information about
the involvement of EEPI in visual localization in the presence of
saccadic eye movements was available from experiments employing a
subjective method of measuring the time course of a saccade in the
nineteenth century before the advent of apparatus for objectively
measuring eye movements was available. These experiments (Lamansky,
1869) employed a brief regularly repetitive flash issuing from a
single physical location in order to ascertain the time course of
the eye movement. From subject's reports of the number of spatially
adjacent flashes observed and the experimenter's knowledge of the
flash rate, some excellent estimates of saccade duration were obtained. These old observations pointed to two important things regarding stability of visual localization:

(a) Stimulation presented during a saccade can be visible.
Since intermittancy of stimulation during a saccade resulted in
seeing the stimulation, the fact that we normally do not see stimulation during saccades must be connected with the continuity of
normal visual stimulation to the retina during saccades.

(b) Cancellation based on EEPI is not perfect at best, and at
worst not present at all. Since the members of the set of images
appeared in adjacent and thus in different locations (even though
they were all generated from a target at a single physical location)
at best only one image could have been visually localized correctly.
(This "intermittent light illusion" is quite readily observed by
saccading across a neon lamp electrically powered by AC against a
dark background; here the lamp normally flashes 120 times/sec. A
series of separate flashes are readily perceived spread out across
a 10° or a 15° dark field bounded by the saccade.)

5.2. Cancellation is Not Needed for Stable Visual Localization in the Presence of Saccades in Structured Visual Fields

As described above (1.2 en 3.2) all visual localization is accurate with steady gaze in any given direction in a structured visual field even though EEPI may be in error and uninvolved in the localization. But if this is so visual localization is accurate at each end of a saccade without EEPI and there is no need for EEPI to be involved in translating local signs to create perceptual stability. Thus, when the eye turns between the two ends of a saccade, a cancellation mechanism for visual localization of stimulation presented in the brief time period of the saccade would be superfluous at best and deleterious to localization at the very least.

Cancellation could serve some purpose for saccades in viewing structured visual fields only if it was necessary to stabilize visual localization of stimulation presented to the eye during a saccade. But stimulation presented to the eye during saccades normally is not visible unless particular experimental arrangements not normally present outside the laboratory are employed. Thus a cancellation mechanism for such stimulation would serve no purpose at all. (As noted below, however, in darkness an EEPI-mediated cancellation mechanism is involved in visual localization for saccades although its spatiotemporal characteristics are crude relative to the precision and accuracy of visual localization in structured visual fields; it is in the latter case that the cancellation mechanism is normally not involved in visual localization.)

5.3. Three Problems for Perceptual Stability in the Presence of Saccades

The fact, that all visual localization is accurate at both termini of a saccade in normal illumination without the involvement of EEPI, is related to the solution by the visual system of three problems in connection with saccades. The problems are: (a) How to eliminate the presaccadic view of the visual field from perception so that its retinal placement will not interfere with the post-saccadic view. (b) How to eliminate from perception the "smeared" retinal information that is generated by the saccade so that it will not interfere with the postsaccadic view. (c) How to maintain a correct correspondence between spatial localizations mediated by vision and by other sense modalities (stability of intersensory localization) and limb placement (stability of sensory/motor localization).

The first two problems ((a) and (b) above) are different but related: The presaccadic view has been seen; it has existed in perception and has to be removed and stored in memory in a way such that localization of stationary objects is the same within memory and within the postsaccadic view. The "during-saccade smear" does not need to be perceived, in fact it should not be perceived; if it were perceived, it could only be done after the saccade itself ter-minated, and it would thus interfere with or delay the perception of the postsaccadic perception. The smear must be eliminated and the stimulus not seen. The third problem appears very different from the first two, but in fact has a great deal in common with them: All three problems require the reconciliation of discrepant information. *The first two are solved by Type B suppression; the third by Type A suppression (Matin, 1981).*

5.4. Type B (Saccadic) Suppression of Visibility

The perception of stimuli presented shortly before and during a saccade is wiped out extremely effectively by postsaccadic stimulation. Since each target "carries its own suppressor" with it, nothing is normally seen of the "smeared" image of stimulation during the saccade (Matin, 1972, 1976b; Matin, Clymer and Matin, 1972; Holly, 1975; see Matin 1974 for review; MacKay, 1970; Brooks and Fuchs, 1975; Mitrani, Mateeff and Yakimoff, 1970a, b; Brooks, Impelman and Lum, 1980). In fact, it is quite important that active suppression be employed since normal persistence of visual stimulation may continue in perception for as much as 300 milliseconds or more following the termination of the stimulation itself (Efron, 1970a, b; Bowen, Pola and Matin, 1974; Matin and Bowen, 1976). Mislocalizations that occur when suppression is removed will be described in the following sections.

Here we note that the main mechanism for Type B suppression ("saccadic suppression") of visibility of any given target is carried out by way of the spatiotemporal sequence of retinal stimulation produced by the target itself and appears to have nothing to do with either EEPI or cancellation per se. This is clearly demonstrated (Matin, 1972, 1976; Matin, Clymer and Matin, 1972; also see Mateeff, 1978, Fig. 3; Holly, 1975) by the fact that a briefly-flashed thin vertical line whose illumination onset is temporally coincident with a horizontal saccade's beginning will appear to have a horizontal extent whose length increases with and corresponds to the extent of retina across which the line's image is distributed by virtue of the eye movement (Fig. 5). But this increase in perceived horizontal extent only holds for increases in flash duration up to the duration of the saccade. If the flash duration extends into the postsaccadic period, the perceived horizontal extent of the smear appears *shorter* than the perceived extent when the flash duration equals the saccade duration, and at moderate flash intensities, increases of flash duration within the postsaccadic period produce still further decreases in the perceived extent, until for a flash duration equal to 100 msec the line width appears no different than either an instantaneous presentation or a normal continuous presentation with no saccade. At lower intensities, although the perceived extent decreases when saccade duration is smaller than flash duration, the smear continues to have a perceived extent that is considerably longer than for continuous presentation even for flash durations of 300 msec. Since the increased duration

of flash beyond the duration of the saccade only adds energy to the
retinal locus of the postsaccadic image on an essentially stationary eye, the decrease in visually perceived length of the smear with
increase in duration must be due to a metacontrast-type of inhibition
produced by this buildup of energy; the larger and more prolonged
the buildup, the greater the suppression of perception of the smear
in adjacent regions. It is thus clear that each target carries its
own suppressor. Interference by other adjacent targets increases
the effectiveness of the masking.

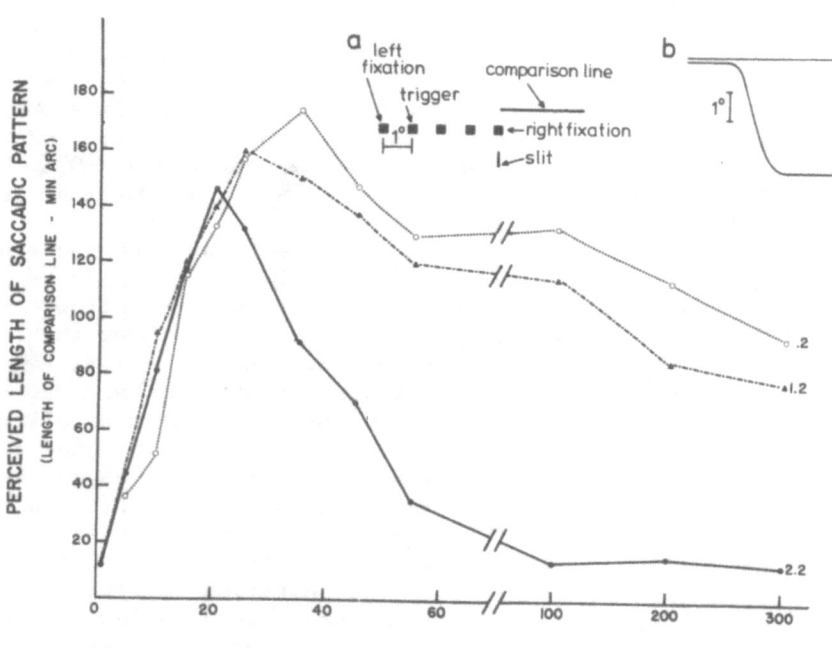

Fig. 5. (From Matin, 1972; Matin, Clymer and Matin, 1972.)
        Perceived horizontal extent of a 2' wide slit for which
        onset of illumination was at the point at which the eye had
        traveled 1° into a 4° horizontal rightgoing saccade (see
        inset a). The slit remained illuminated for the duration
        indicated by the value on the abscissa. The three values of
        slit luminance are shown to the right of the curves in log
        ml. Inset b is a recording of a change in eye position
        during a 4° saccade.

The view that the Type B suppression in Fig. 5 is a consequence
of the spatiotemporal pattern of retinal stimulation is strongly
supported by the experiments of MacKay, 1970; Mitrani, Mateeff and
Yakimoff, 1970a, b; Brooks and Fuchs, 1975; Brooks, Impelman and
Lum, 1980. Thus MacKay (1970) found that with a steadily fixating
eye an instantaneous movement of a 10° circular background produced
an elevation in visual threshold for a test flash in the center of
the background. The elevation follows the same time course as does
the threshold elevation in the presence of a voluntary saccade. It
begins 50-100 msec prior to the beginning of the saccade, rises to a
maximum during the saccade, and decreases monotonically until 200-
300 msec after the saccade when it approximates the value of threshold temporally distant from the saccade. The magnitude of threshold
elevation in the presence of saccadic eye movements is influenced by
the temporal, spatial, and pattern relations between test flash and
background (Mitrani, Mateeff and Yakimoff, 1970a, b; Brooks and
Fuchs, 1975).

The major component of Type B suppression described above
appears to be capable of eliminating from perception smears that
are many log units above threshold. The interference caused by the
spatiotemporal pattern of retinal stimulation can be broken into a
number of types and several sources that are only indicated by the
brief description above. The description here will not follow this
delineation. But there is a second much smaller component of saccadic
suppression that has been observed in ways that are not ascribable
to masking from moving or stationary backgrounds or fixation points,
or to variations in retinal sensitivity across retinal regions (cf;
Pearce and Porter, 1970; Greenhouse and Cohn, 1980; Greenhouse,
1981; Volkmann, 1962; Volkmann, Schick and Riggs, 1968; Zuber and
Stark, 1966; Riggs, Merton and Morton, 1974). By a signal detection
approach Pearce and Porter (1970) established the existence of a
"criterion-free" change in sensitivity as well as a criterion change
whose time course paralleled the time course of the threshold change.
Since trial-to-trial variability exists for the perceived spatial
location of flashes presented during saccades (Matin and Pearce,
1965; Matin et al., 1969), it was suggested that the increase in
threshold for flashes in darkness during a saccade is a consequence
of increased spatial uncertainty just as introducing spatial uncertainty for measurements of threshold with a fixated eye also
raises threshold (L. Matin, quoted in E. Matin, 1974, p. 910). The
actual existence of this component and its basis in spatial uncertainty has been demonstrated by showing that a clearly visible
pedestal for the test flash which marks spatial location also
eliminates the suppression (Greenhouse, 1981; Greenhouse, Cohn and
Stark, 1977; Greenhouse and Cohn, 1980). Although the proof of the
existence of an influence of spatial uncertainty does not prove the
nonexistence of other central effects proposed as explanations for
the .5 (or so) log units of suppression that can remain after
factors related to the spatiotemporal pattern of retinal stimulation are eliminated, it does suggest they are less likely. Thus,

for example, if spatial uncertainty exists for phosphenes produced during saccades (as is extremely likely) influences of EEPI in reducing the quantity of an excitatory neural response to a light flash are not required to explain the results of Riggs, Merton and Morton (1974) who observed a suppression equivalent to about .4 log units for electrical phosphenes.

5.5. A Failure of Cancellation Theory: Saccades

Four separate sets of experiments quite clearly demonstrate that cancellation does not work for stimulation during saccades in the way it would need to if it were the primary mechanism for stabilization of visual localization for stimulation during saccades. One of them will be described here. The others will be described in the next three sections:

Two vertical lines of a vernier target were simultaneously presented above each other in a brief 1 msec flash on the central fovea (Matin, 197b). Presentation was at the moment the eye was in the center of a voluntary horizontal 4°13' saccade carried out in darkness following removal of both the fixation target and saccadic goal. The salient point about this experiment is that both precision and accuracy of the vernier discrimination were unchanged regardless of whether the top line was two log units brighter than the bottom line, or the bottom line was two log units brighter than the top line, or the intensities of both lines were equal. Since onset latency of response in the visual system decreases with intensity and since a large portion of this decrease occurs in the retina, the message from the brighter line reaches the optic nerve earlier. Any cancellation mechanism involving an influence of EEPI on a message derived from the optic nerve would be expected to produce a perceived displacement between the two lines corresponding to the distance the eye traveled in a time equal to the latency difference. An offset of roughly 12'/msec of latency difference is expected on this basis. No perceived displacement was obtained. An offset of even 1' would have been easily discernible experimentally and at least a 10 msec latency difference would be expected (yielding 120' of offset). Thus one must conclude that cancellation by EEPI was not involved. Of course, one could ask why a "smart" cancellation mechanism that could also compensate for the intensity-difference-produced latency-difference could not be inferred. But for any such mechanism to work sufficiently well to yield perceptual stability as is normally observed in normal illumination, information regarding time of arrival of quanta at the photoreceptors would need to be much more precise and also be available independently of the stimulus striking the eye itself; we can be quite sure that the system is not that smart. (See section on visual persistence below (and Fig. 7) for analogous results where time between two flashes presented during a saccade was manipulated directly.)

5.6. Spatiotemporal Characteristics of EEPI Used for Visual Localization in the Presence of Saccades in Darkness

The experimental paradigm of Fig. 6 has been employed to determine the dependence on eye position and time course of EEPI as it is involved in visual localization in the presence of voluntary horizontal saccades. In this paradigm the test flash was presented to a known retinal locus in the eye of an observer at a known time. The observer reported where the test flash appeared ("left" or "right") relative to a fixation target that had been visible earlier but was extinguished several hundred milliseconds prior to the saccade. Except for the few briefly-presented stimuli involved in the measurement procedure the subject was in darkness.

Fig. 6a shows on the ordinate both (a) the average location at which the test flash had to be placed in order to be perceived as lying in the same horizontal visual direction as the fixation target (TPSE = Target PSE), and (b) the average eye position (EP). Both of these are plotted against time from the beginning of the saccade for each of two attempted saccade lengths: 5° (5 NA) and 8° (8 NA). (The 8A condition will be described below in section 5.10.) For both saccade lengths and at all times displayed on the abscissa the eye undershot the saccade's goal (EP measurements) and the subject mislocalized the previously-seen fixation target as measured by the flash (TPSE measurements) by from 1° to 2° for the 5° saccades and from 2.5° to 3° for the 8° saccades. While the subject shown in Fig. 6 mislocalized the fixation target in a direction opposite to the saccade, this is not universal. For example, some other subjects typically mislocalize the previously-seen fixation target by as much as 5° in the direction of the saccade. These substantial errors in localization measured at the target array involve substantial errors at the retina. In Fig. 6b average errors in Retinal PSE reach somewhat more than 1° at 200 msec; for other subjects errors in Retinal PSE as much as 7° are obtained as early as 15 msec into an 8° saccade. Similar results had previously been obtained for 2° 13' and 4° 26' saccades (see Matin, 1972, 1976b). Clearly the errors are a consequence of the occurrence of the saccadic eye movement (an error in EEPI and of cancellation). These errors are not a consequence of memory failure for the location of the visual target as Skavenski (1976) has claimed. They are too large. Direct determination of the memory-error component in isolation yield values that are not more than 15' (Matin et al., 1981a; see section 3.1.3 above).

If the normal perfect constancy of visual direction was solely the outcome of a cancellation mechanism, such cancellation would have to make use of EEPI in a way (a) so that the time course of the change of the Retinal PSE would exactly mirror the time course of the saccade, and (b) so that visual localization is accurate always in the experiments described above. The failure of both (a) and (b), along with other errors of visual localization for stimulation before a saccade (Matin, Matin and Pola, 1970), implies (for reasons other than those pointed to earlier) that cancellation cannot be

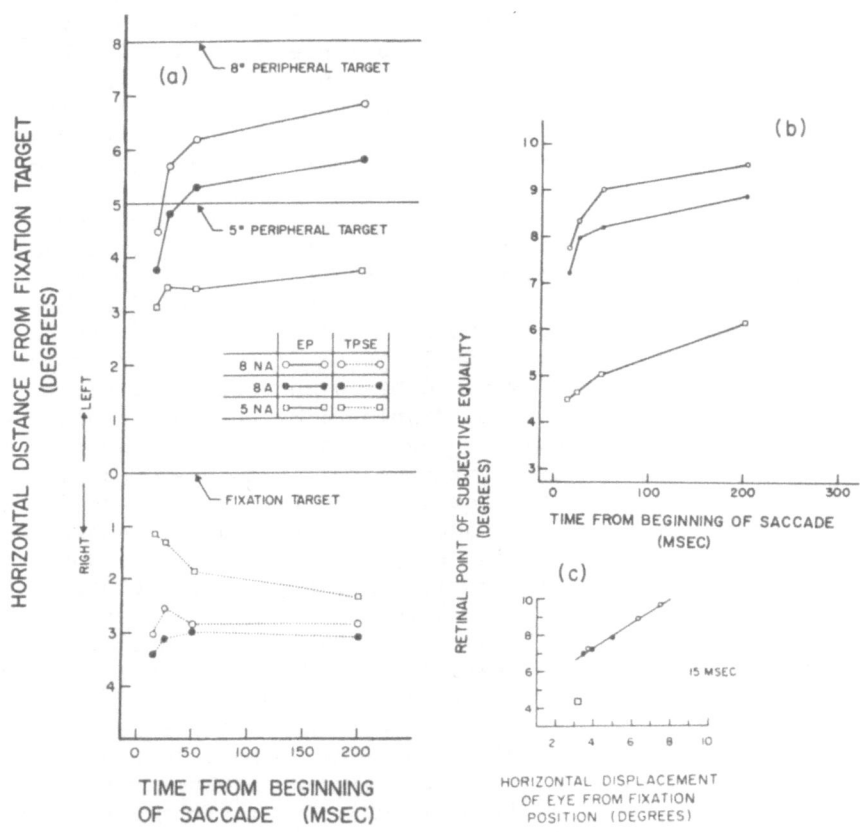

Fig. 6. (a) Eye position (EP) and Target PSE (TPSE) during and
after a 5° (□) and 8° (O) saccade as a function of time;
(b) Retinal PSE as a function of time;
(c) Retinal PSE at 15 msec as a function of eye position.
Results with NA (nonadapted) conditions described in Section
5.6.; results with adapted (parametric adjustment) condition
8A (filled circles) described in Section 5.10. (From Pola,
1972.)

responsible for the high degree of accuracy of the normal constancy of visual direction we experience in the presence of saccades in normal illumination.

Figure 6c does show that cancellation can and does contribute to visual localization in darkness in an important way. This figure is a breakdown into groups of the data for each condition in Figs. 6a and 6b at 15 msec into the saccade. The groups differ along an eye position axis. For trials in which the eye had moved further from the original fixation position the Retinal PSE was also larger, indicating an increase in the EEPI-caused shift in visual localization. The increase in Fig. 6c for the 8A condition is linear with the increase in eye position. A similar linear increase was obtained at each of the durations after the saccade beginning. Three important conclusions follow from this results:

(a) The variation of Retinal PSE with eye position must be the expression of an EEPI-based cancellation mechanism;

(b) The mechanism is sloppy relative to the precision of visual localization when the comparison of locations is for simultaneously-present objects. The sloppiness is expressed in several ways: (1) Standard deviations of perceived location are typically about 1°; (2) The variation of Retinal PSE with eye position as in Fig. 6a has been found to have a slope between .4 and .76 indicating much less shift in Retinal PSE than in eye position; (3) The actual error in localization can be as much as 5° for an 8° saccade; (4) The time course of the shift of Retinal FSE does not mirror the shift of eye position. Much larger departures were obtained with other subjects and other saccade lengths than are displayed in Fig. 6.

Since the subject attempted to carry out the same length saccade on each trial, variation in actual saccade length was presumably unintentional; the effect in Fig. 6c, however, shows that perceived visual localization was established with the help of information about saccadic length. It is thus very likely that at least some aspect of this information was inflowing.

5.7. Influence of Visual Persistence on Visual Localization in the Presence of Saccades

A central feature of the paradigm described in the previous section is that the test flash was presented in darkness several hundred milliseconds following any previous visual stimulation, and that generally at least 1/2 second intervened between viewing the two items whose visual localizations were compared. The visual direction report for the test flash was thus made relative to the memory of the previously-viewed fixation target. This 1/2 second minimum interval was originally employed to deal with the fact that visual persistence of even a 1 msec visual stimulus is considerable and would extend the duration of perception of the flash to a value between 100 and 300 msec (Bowen et al., 1974; Matin and Bowen, 1976). Since the object of those experiments was to plot the time course of EEPI involved in visual localization, the experimental plan re-

quired eliminating factors that might force relative localization
to be judged simply on the basis of the relative retinal locations
of the stimuli without any attention to the change in eye position.
If the visual perception of the first of two stimuli whose relative
localizations were being compared persisted into the time period
during which the second stimulus was visually perceived it was
thought that the likelihood that relative retinal location would be
the main basis for the judgment would be increased; any EEPI involved in a cancellation mechanism might then be expected to operate
similarly on the two stimuli, and the time course of EEPI would not
have been what was measured.

Subsequent experiments specifically designed to yield measurements of the influence of visual persistence on visual localization
directly bore out this concern and also showed that the earlier
experiments had taken the appropriate precautions (Matin, Pola,
Matin and Bowen, 1971; Matin, 1976b). In Fig. 7 two sets of data are
shown for each subject:

(a) The open circles are results obtained in the same way as
the results in the previous section: The location of a test flash
that was presented at a time following the beginning of a saccade
(abscissa) was compared by the observer with the location of the
previously-viewed fixation target that had been extinguished about
3/4 sec before time zero on the abscissa. Saccade length was 8° for
one subject (upper panel), 5° for the second subject (lower panel).
(The difference between the two subjects in speed of the change of
retinal PSE at short times is characteristic for them.)

(b) The solid-circle results are from a subsequent experiment
employing the same paradigm as in (a) with one major exception: In
this case the subject compared the relative locations of two 1 msec
flashes. The first of the two (standard flash) was presented when
the eye had reached 1° into the saccade and was presented to the
same retinal location on all trials. That location was the one that
had been found to be the Retinal PSE for the fixation target in the
previous experiment whose results are shown by the open circles.
The second flash was presented at a specific time after this first
flash, and to a retinal position that was randomly varied from
trial to trial. The Retinal PSE for the first flash was measured
with this second flash. Since the second flash in this experiment
(the test flash) occurred at the same time relative to the saccade
as the test flash in the first experiment, the variation of Retinal
PSE with time should be identical for both experiments if this
variation was solely a function of the operation of a cancellation
mechanism. But there is a large divergence between the two sets of
data for each subject in the early times (from 0 to 200 msec for
one subject; from 0 to 100 msec for the second subject) followed by
reasonably close convergence only at the longer times. This attests
to an important difference in the mechanism controlling visual
localization in the two experiments at the shorter time intervals.

The importance of this difference and its basis in visual persistence can be seen by considering one part of the results. Consider

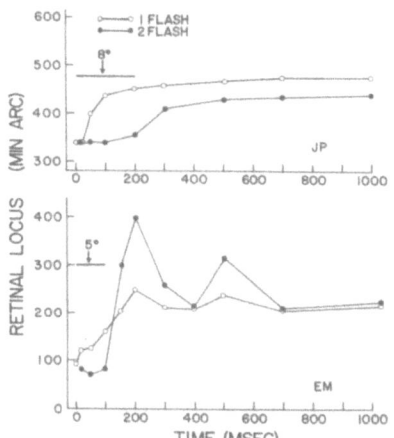

Fig. 7. (From Matin, Pola and Matin, 1972; Matin, 1976b.)
The experiment shows the influence of visual persistence on
visual localization in the presence of saccades.
Open circles (O) are results (Retinal PSE values) when the
observer compared the location of a 1 msec test flash to
the location of a previously-extinguished fixation target.
The abscissa is time after the observer's eye had reached
1° into the saccade.
Solid circles (●) are Retinal PSEs of a second flash for
the location of a first flash. The first flash was presented
when the eye had reached 1° into the saccade. The location
of the first flash was fixed at a position that was observed
to be the Retinal PSE for the fixation target (as measured
in the experiment of the previous paragraph). The subject in
the upper figure attempted an 8° saccade; a different subject
in the lower figure attempted a 5° saccade.

the upper panel: 100 msec after the beginning of the saccade the
Retinal PSE for the fixation target had already departed by about
110' from its value at 0 msec (change from 340' to 450', ordinate
values, open-circle values, Fig. 7). Yet when the standard was presented as a flash at 0 msec, the Retinal PSE for this standard as
measured by flashes 100 msec later had barely moved from the Retinal
PSE at 0 msec (solid-circle values, Fig. 7). Thus, stated another
way: when two flashes were presented on a given trial, one each at
0 and 100 msec, they had to be presented at almost the same retinal
location in order to appear in the same location, although the
Retinal PSEs for the previously-viewed fixation target were very
different at those same two times. Two important inferences follow
from this result: (1) *Transitivity has failed:* two Retinal PSEs
taken separately for the previously-viewed fixation target (one at
0 and a second at 100 msec; open-circle results) are not equal to
each other when directly compared (solid-circle results). (2) *The
Transitivity failure took place in just the region of time over which
we would expect persistence for the first flash to exist*, suggesting that it was the simultaneity of perception of the two flashes
in the closed-circle results that led to the Retinal PSE holding
at a nearly fixed value for the interflash intervals up to 100-200
msec.

Along with other results in Section 5, these results establish
constraints by retinal processes on the operation of an EEPI-guided
cancellation mechanism. They also imply the need for Type B suppression to clear out persisting effects of earlier stimulation if
retinal direction values ("local signs") are to be rapidly modified
by cancellation. The results in this section demonstrate the potency
of one set of retinal constraints on visual localization. Another
aspect of this matter is dealt with in the next section on the influence of background on visual localization.

5.8. Influences of Visual Background on Visual Localization in the Presence of Saccades

The experiments described in the two previous sections involved
presentations of test flashes in darkness at least several hundred
milliseconds away from all other visual stimulation. A measure of
the influence of other visual stimulation located more closely in
time to the test flash is given by the result for short time intervals in the last section. But most of our visual behavior is not
carried out in darkness. It is carried out in illumination which
provides a visual background of contours, different brightnesses,
and objects against which any single visual object may be localized.

Several workers have carried out experiments in which flashes
presented before, during, or after saccades are localized relative
to objects on a steady background (Bischof and Kramer, 1968; Matin,
Matin, Pola and Kowal, 1969; Mateeff, 1978). All subjects manifest
very substantial and frequent localization errors during and after
the completion of the saccade, with the largest reported (but not

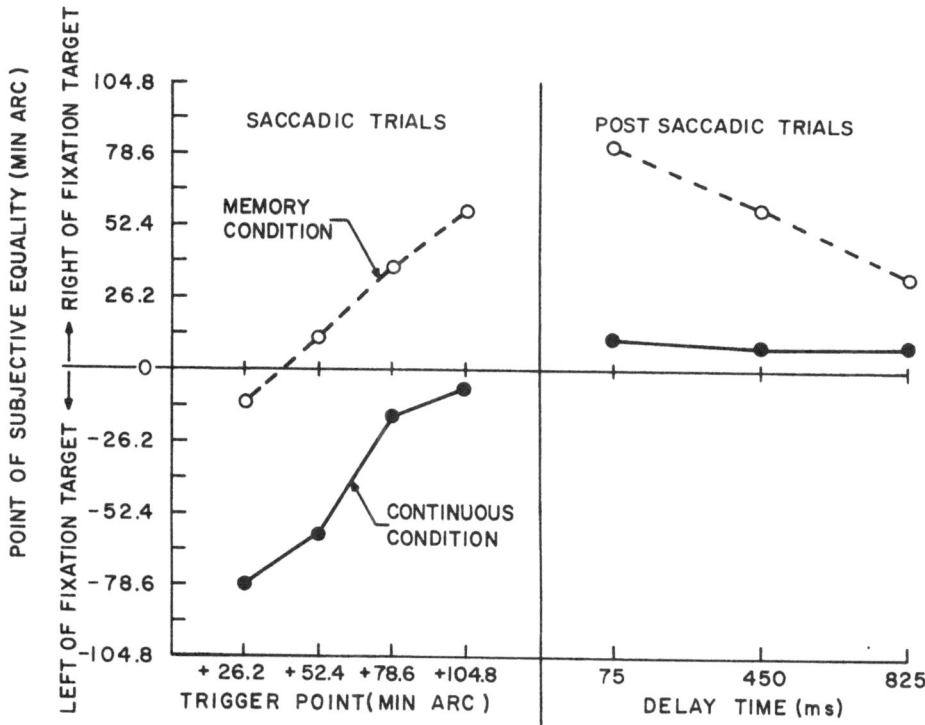

Fig. 8. (From Matin, Matin, Pola and Kowal, 1969; Matin, 1976a.)
Visual localization measured with test flashes presented
during 2° 13' rightgoing saccade (left half of figure
("saccadic trials")) and after saccades (right half of
figure ("post saccadic trials")) in the presence of a
steady background ("continuous condition") and in total
darkness ("memory condition"). The entire steady background
consisted of a single point of light located .5° above the
fixation target. Trigger point (on abscissa) refers to eye
position (as a distance from the fixation target) at which
the 2 msec randomly-located test flash was presented during
the saccade; delay time refers to time after completion of
the saccade. See text for further description.

at all untypical) being 15° during 16° saccades (Bischof and Kramer,
1968). A very important difference between localization against a
visible background and localization of a flash in darkness relative
to a previously-viewed target is that, for flashes against a background the errors are reduced to very small values rapidly after the
saccade, whereas in darkness they may remain high for some period
(Fig. 8). In effect, against the background the test flash presented
shortly after the saccade must be presented at very nearly the same
retinal location as the target with which its location is being com-

pared for it to appear in the same visual direction. These results
are readily explicable by the Dual-Suppression theory: (a) Type B
suppression operates on the steady background so that it is seen
shortly after the saccade with presaccadic and during-saccade views
eliminated. (b) This postsaccadic view of the background determines
the postsaccadic coordination of retinal locus and perceived location (retinal local sign). (c) Type A suppression from this background prevents the EEPI-based cancellation mechanism from determining localization of postsaccadic test flashes; localization of postsaccadic test flashes is based solely on the relation of the locations of their retinal images to the retinal images of inhomogeneities in the continuously present background, which themselves are
normally treated as stationary. (d) This influence of the background
on visual localization extends ("backward") to flashes presented
during the saccade. Those flashes are easily visible since the conditions have been arranged to prevent Type B suppression. Like the
flashes presented after the saccade, the test flashes presented
during the saccade also appear to be in the same location as the
fixation target only when they strike very nearly the same retinal
location as does the fixation target presented postsaccadically. For
them too Type A suppression has set in (a "backward" influence).
*It is of some importance to note that the entire steady background
that produced this time-binding across pre- and postsaccadic views
(and Type A suppression) was a single 3.5' circular target.*

5.9. Type B (Saccadic) Suppression for Perception of Displacement

In addition to visual mislocalizations, increases in trial-to-
trial variability in visual localization, and the reduction in detectability of targets which accompanies voluntary saccadic eye
movements and which have been described above, a number of workers
have reported a reduced sensitivity to spatial displacements and
movement (Ditchburn, 1955; Wallach and Lewis, 1966; Bridgeman,
Hendry and Stark, 1975; Stark, Kong, Hendry and Bridgeman, 1976;
Mack, Fendrich and Pleune, 1978; Festinger and Holtzman, 1978;
Whipple and Wallach, 1978; Bridgeman and Stark, 1979; Heywood and
Churcher, 1981). This is measured by an increase in threshold for
the detection of movement or displacement (reduced discrimination
between movement and no movement) or reduced discrimination between
movement in different directions. The central (but not exclusive)
interest in these phenomena arises from the possibility that it is
a manifestation of an EEPI-mediated cancellation mechanism. However,
to what degree such a mechanism is involved is not yet clear.
     In almost all cases the reduction in sensitivity has been for
movement of the entire visual field (whether that field was a single
point or a complexly-structured field). A similar result had also
been obtained with involuntary saccades (Beeler, 1967). It is important to contrast this reduction with the facts noted above that
sensitivity for vernier discrimination during either voluntary
(Matin, 1976b) or involuntary (Krauskopf et al., 1966) saccades is

no different than when no saccade is present: The vernier discrimination involves discrimination of displacement between two simultaneously-present lines; the result presently under discussion is a discrimination of a field's instantaneous change in position from a previous position in darkness. It is not yet clear whether the effects under discussion are unique to saccadic eye movements, whether they have any bearing on the involvement of EEPI in spatial localization or changes in spatial localization, or to what degree the effects depend on purely retinal processes (see Glossary) including effects of stimulus complexity, retinal eccentricity of the critical stimulus, and metacontrast or other masking effects related to the saccade.

One of the aspects that might be unique to eye movements, although not necessarily to saccades, is the possibility of confusability between the shift of the retinal stimulus resulting from the eye movement and the retinal shift resulting from the experimenter-produced shift of the visual field. The idea of such confusability as the basis for the effect is supported by (1) an increase in threshold proportional to saccade length and (2) a threshold increase of stimulus displacement in the direction orthogonal to the saccade. However, although both results have been reported in some of the reports listed above, others have failed to agree on the matter. Another major difference among reports is the wide range of reported thresholds: threshold values for target motion as large as the eye movement itself (Wallach and Lewis, 1966) and as small as 3' for displacement during a 10° saccade (threshold = 1/200 of saccade distance) (Stark et al., 1976) have been reported. Although some of the wide range in threshold magnitude in the results of previous studies can be explained by large differences among subjects in criteria with identical d' as has been found recently (Matin and Rogan, unpublished data), this will not acount for the entire range. Some portion of the variation in threshold magnitude is certainly also a result of differences among stimulus conditions employed in the different experiments. Both the time at which the experimental stimulus displacement has been introduced relative to the saccade and the actual time course of the stimulus displacement have differed among the different studies. The loss in detectability of stimulus displacement does depend very substantially on these variables and follows the same time course as does the loss in detectability of the presence or absence of a flash in the presence of a saccade (Bridgeman et al., 1975).

5.10. Parametric Adjustment and Visual Localization

A very important example of the separability of visual and motor localizations lies in the presence of parametric adjustment (McLaughlin, 1967; Pola, 1972, 1976; Miller, 1980). On each trial an individual first fixates one target (A) that is horizontally displaced from the only other visible target (B) in an otherwise dark field. At a prearranged signal, the observer saccades to B;

during the saccade B is extinguished and simultaneously replaced by
a target B' that is horizontally displaced from B and closer to A.
Thus the goal to which the saccade is directed is not present when
the eye arrives at the end of the saccade. On the first trials employing this procedure the eye reaches B but "reflexively" carries
out a second saccade to B' after a short latency. On subsequent
trials the primary saccade length is reduced until only a single
saccade to B' is carried out. This ocular adjustment is not a conscious correction. In fact the observer is not aware of the fact
that B has been replaced by B'. Instead, he reports a "jumping" of
the visually perceived saccadic goal on the initial trials at a time
which corresponds in time to the secondary saccade (even though he
may not realize that he carried out such a second saccade); the
appearance of "jumping" is completely eliminated from perception
when parametric adjustment is complete; at this latter stage the
observer reports that he has reached the saccadic goal and that the
appearance of the events and visual stimuli are no different than
when he was not adapted and no B-B' switching was employed (i.e.,
the appearance is "normal"). Parametric adjustment has two immediate
implications for us here: (1) Following the attainment of parametric
adjustment, B as seen before the saccade appears in the same visual
direction as B' viewed after the saccade although the two may be
separated by a substantial distance (e.g. separation by 3° for the
case where the A-B distance is 8°). This suggests that perceptual
and motor adjustments occur together with regard to the saccade's
goal. In fact this is so with regard to the appearance of B. (2)
However, in an experiment in which parametric adjustment was employed to reduce saccade length and in which the observer reported
on the relation of visual direction of a subsequent TF to the
original fixation target (condition 8A, Fig. 6), the relation of
Retinal PSE to eye position was indistinguishable for the condition
(8NA) in which parametric adjustment had not occured (Fig. 6c). This
result is further strong support for the involvement of an inflowing signal in visual localization.

6. CONCLUDING REMARKS

6.1. Unfinished Business

Most of the work described above involved a perceptual measure
of localization. There are two main ways in which this treatment is
incomplete in dealing with presently available information: (1) Localization during steady gaze and saccades have been dealt with but
localization in the presence of pursuit eye movements, torsional
eye movements, and disjunctive eye movements have not. (2) It would
be desirable to describe additional work in which the localizing
response is itself a motor response such as (a) when subjects turn
their eyes (motor) to foveate a nonfoveally-presented visual target
(sensory), (b) turn their eyes (motor) to fixate a point in space at

which an unseen sound (sensory) is located, or (c) position an arm
or a held object (motor) to the location at which a visual target
(sensory) is presented. These matters will be dealt with and related
to the picture derived from the work described in this chapter in
another chapter that is in preparation and will be presented elsewhere.

It will be worth listing some further references on these
matters although constraints of time and space do not permit going
further: on foveating a nonfoveally-presented visual target
(McLaughlin, 1967; Pola, 1972, 1976; Shebilske, 1976; Hallett, 1976,
1978; Hallett and Lightstone, 1976a, b; Hansen and Skavenski, 1977;
Hansen, 1978; Becker, 1976; Becker and Klein, 1973), on turning the
eyes to an unseen sound (Zahn, Abel and Dell'Osso, 1978), on positioning an arm to a visual target (Merton, 1961; Hansen and
Skavenski, 1977; Kornheiser, 1976; Welch, 1978), on visual localization and the visual perception of movements in the presence of
pursuit movements (Stoper, 1973; Ward, 1976; Sedwick and Festinger,
1976; Wertheim, 1981), on visual localization in the presence of
torsional eye movements (Nakayama and Balliet, 1977; Balliet and
Nakayama, 1978a, b) and on the possible involvement of EEPI in the
presence of disjunctive eye movements.

6.2. Structured Visual Fields, Sight of One's Own Body, and Differ-
     ences Between Visual Localization in the Horizontal and Verti-
     cal Dimensions

A large part of this chapter has been devoted to concerns
relating to cancellation mechanisms – not so much the source of
EEPI as the necessity and sufficiency of cancellation for explaining various aspects of spatial localization, and particularly visual
localization.

The central observation on this matter with partially paralyzed
observers (that all visual localization during steady gaze in normally illuminated and structured visual fields is accurate although
EEPI manifests huge errors) was made with no attempt to prevent the
observer from viewing his own body. Indeed, sight of one's own body
is a normal part of one's view of the world and so these observations do pertain to the broadest range of normal viewing conditions.
The important inference from these observations is that cancellation
does not influence visual localization under these conditions.

Experiments are in progress in which we expect to completely
eliminate sight of the partially paralyzed observer's body while
visual localization is again measured. It is expected that substantial Type A errors in visual localization will be obtained within
the horizontal dimension, (i.e., in judging the location of a light
relative to the observer's visually perceived median plane) but that
no errors in visual localization of a light will be observed within
the vertical dimension (i.e., in judging the location of a light
relative to the visually perceived horizontal).

The basis for these expectations is that physical space is not

polarized along the horizontal as it is along the vertical (gravity);
consequently localization within the horizontal dimension cannot be
made with respect to any intrinsically ordered property of physical
space, but can only be made relative to the observer's own body.
(This assumes of course that visual inhomogeneities or objects are
not present in the observer's visual field to "intrinsically polar-
ize" the dimension.) Removal of the sight of his own body in an
otherwise normally illuminated and structured environment thus re-
moves the unique visual reference for localizing a light relative
to the observer's visually perceived median plane and consequently
should bring the EEPI-mediated cancellation out of suppression and
into full play (with only visual memory of the relation of his body
to the inhomogeneities in the environment as a possible additional
contributor.) Hence it is expected that with sight of his own body
removed the partially paralyzed observer will err in judging the
location of a light relative to his median plane and that these
errors will be exactly the same as his errors in total darkness.

However, directions from the observer's eye along the vertical
(up-down) are uniquely specifiable with reference to the direction
of gravity. Horizontal is thus a physically unique direction whose
actual direction does not depend at all on the orientation of the
observer's body in physical space. Visual perception of the orientation of his own body should thus not normally assist an observer in
specifying the eye-level horizontal. However, since one of the two
main sets of lines of organization of structured visual fields normally coincides with the direction of gravity and accurately speci-
fies the direction of the vertical, an observer could accurately
specify eye-level horizontal by referencing his judgment to these
main lines of the visual field. The partially-paralyzed observer
accurately localizes the eye-level horizontal in normal illumination
but not in darkness. This suggests that he does indeed make use of
the main lines of organization of his visual field for perception of
eye-level horizontal, and that when this visual field is visually
perceived EEPI-mediated cancellation is suppressed. These results
also suggest that if the (paralyzed or normal) observer's room is
tilted within his median plane, errors in setting a light to the
perceived eye-level horizontal will occur and the setting will be in
the direction specified by the main lines of organization of the
tilted room; such a result would converge with results with rooms
that are tilted around the horizontal line of sight of the observer
(Witkin and Asch, 1948; Witkin, 1949), and would suggest a convergence of mechanisms for Type A suppression in the two directions.

This approach then suggests that sight of the observer's own
body without sight of the rest of his visual field leaves Type A
suppression operative for judgments of visual localization relative
to the median plane, but allows EEPI-mediated cancellation to function unsuppressed for visual localization of the eye-level horizontal, while sight of the main lines of organization of the visual
field without sight of his own body should leave Type A suppression
operative for visual localization of perceived eye-level horizontal

but allow EEPI-mediated cancellation to function unsuppressed for
visual localization relative to the observer's median plane.

The predictions made above for visual localization by the
normal or partially paralyzed observer in an illuminated visual
field with or without sight of his own body and with or without room
tilt, assumed an observer whose bilateral symmetry was set within a
vertical plane. If the observer himself is set to the horizontal
(lying on his back looking upwards) instead of viewing from an erect
position the present formulation predicts that when observing in a
normally illuminated room without sight of his own body, perceived
eye-level judgments along an observer's median plane should be in
error by exactly the same amount as in darkness (since now the EEPI-
mediated cancellation mechanism should have full play), but perceived
eye-level horizontal judgements (along the physical vertical) should
be accurate.

The predictions from the Dual-Suppression Theory of spatial
localization given in this section are presented as part of the
nucleus of a coherent explanation of the observations of the partially paralyzed observer described earlier in addition to their
role as predictions for our ongoing experiments. The basis for presenting them here is that they do suggest a broader set of interactions than we have yet explored between EEPI-mediated cancellation
and Type A suppression as guiding our spatial localization·and
orientation.

FOOTNOTES

[1] It is clear that a setting of a visual target to eye-level
horizontal in darkness depends in part on information regarding the
direction of action of gravity. Since direct gravitational information does not enter through the eyes the question might be raised
regarding whether the setting of a visual target to a visual norm
does not involve an intersensory comparison. Simplicity (and not
logic) suggests that the norm be assigned to the sense modality on
the basis of the phenomological label given it by the observer. On
these grounds when an observer is asked to set a visually presented
line so that it looks horizontal, the norm against which the observer normally and naturally sets the visual target is *defined* to
be a visual norm. Similar comments apply to other visual norms that
might have nonvisual determinants.

[2] Although the motor behavior in sensory/motor localizations
may be under control by sensory information other than that sensory
event to which the match is attempted (e.g., muscle spindle or
tendon organ information determining arm position when a subject
aligns his unseen arm with a light), it is important to emphasize
that it is the motor event which is set to match the visual event.
By classifying sensory/motor localization matches separately from
visual localizations or intersensory localizations, we are able to

deal with the questions at issue without getting simultaneously en-
tangled in the important but separable issue regarding what sensory
and non-sensory inputs control the motor response itself. Thus, just
as in footnote 1 where the visual norm was treated as "visual" regardless of the inputs controlling the norm, here the motor event is
treated as "motor" regardless of the sensory inputs controlling the
motor event. In a word, the terminology is intended to maintain
focus on the output of a "final common path" in both cases.

[3] Although the presentation of the hybrid model (Matin, 1972,
1976a) was intended to deal with phenomena that were previously only
accounted for by outflow theory (and thus to point out that the
inflow-outflow controversy was not closed in favor of outflow
theory), it was not intended as a complete theory of visual stability. Thus, for example, in 1972 I stated: "The evidence presented
... above makes it clear that the extraretinal signal, regardless
of its source, is not sufficient to account for the normal stability
of stationary objects. Nevertheless, the basis for the extraretinal
signal remains a problem (Matin, 1972, p. 368)". The insufficiency
of cancellation theories derived from the previous experiments
mapping the spatiotemporal variation of EEPI in connection with
saccades. This insufficiency was already clear from the first report
of these experiments (Matin and Pearce, 1965) where we were already
able to state: "Although the data from the present experiments (as
well as data from these other studies) do not yet permit a full
picture of the situation, we suggest that proprioceptive compensation is subject to important restrictions related to the temporal
and spatial characteristics of the visual presentation and is more
limited than heretofore thought. We also suggest that a sizable
portion of the normal stability of visual directions during eye
movements results from purely visual factors such as continuously
stable relative retinal positions of stationary objects and from the
continuity of movement across the retina of the images of stationary
objects during eye movements (both of these factors serving to mask
the frequent transient distortions resulting from the brief voluntary saccades which would otherwise be noticeable". Subsequent work
has borne out these suggestions (see below).

[4] The direction of gaze with respect to the head was maintained
at a fixed value by appropriately changing the vertical position of
the fixation target when changes in head-and-body tilt were introduced.

[5] Further confirmation that the errors in setting the test
light to the eye-level horizontal were not a consequence of misper-
ceptions of body tilt lies in the fact that when the subject points
his finger at the horizontal, he does so as accurately in darkness
in the curarized as in the noncurarized state, a level of accuracy
that does not noticeably change with head-and-body tilt. However,
when pointing at a truly horizontal light in darkness under gaze

conditions for which a substantial error is made in setting a
visual target to the eye-level horizontal, the subject points his
finger towards the floor (when he reports the visual target near
the floor) or towards the ceiling (when he reports the visual target
near the ceiling); in normal illumination he points accurately at
the truly horizontal light (Table I).

REFERENCES

Allik, J., Rauk, M. and Luuk, A., Control and sense of eye movement
    behind closed eyelids. Perception, 1981, 10, 39-51.
Balliet, R. and Nakayama, K., Training of voluntary torsion.
    Investigative Ophthalmologyy and Visual Science, 1978a, 17,
    303-314.
Balliet, R. and Nakayama, K., Egocentric orientation is influenced
    by trained voluntary cyclorotary eye movements. Nature, 1978b,
    275, 214-216.
Becker, W., Do correction saccades depend exclusively on retinal
    feedback? A note on the possible role of non-retinal feedback.
    Vision Research, 1976, 16, 425-427.
Becker, W. and Fuchs, A.G., Further properties of the human saccadic
    system: eye movements and correction saccades with without
    visual fixation points. Vision Research, 1969, 9, 1247-1257.
Becker, W. and Klein, H.M., Accuracy of saccadic eye movements and
    maintenance of eccentric eye position in the dark. Vision
    Research, 1973, 13, 1021-1034.
Beeler, J.W., Visual threshold changes resulting from spontaneous
    saccadic eye movements. Vision Research, 1967, 7, 769-775.
Bertelson, P.B. and Radeau, M., Ventriloquism, sensory interaction,
    and response bias: Remarks on the paper by Choe, Welch,
    Filford and Juola. Perception and Psychophysics, 1976, 19,
    531-535.
Bischof, N. and Kramer, E., Untersuchungen und Ueberlegungen zur
    Richtungswahrnehmung bei willkurlichen sakkadischen Augen-
    bewegungen. Psychol. Forsch., 1968, 32, 185-218.
Bowen, R.W., Pola, J. and Matin, L., Visual persistence: effects of
    flash luminance, duration and energy. Vision Research, 1974,
    14, 295-303.
Bridgeman, B., Hendry, D. and Stark, L., Failure to detect displace-
    ment of the visual world during saccadic eye movements.
    Vision Research, 1975, 15, 719-722.
Bridgeman, B. and Stark, L., Omnidirectional increase in threshold
    for image shifts during saccadic eye movements. Perception
    and Psychophysics, 1979, 25, 241-243.
Brindley, G.S., Goodwin, G.M., Kulikowski, J.J. and Leighton, D.,
    Stability of vision with a paralysed eye. Journal of Physiology,
    1976, 258, 65p-66p.
Brindley, G.S. and Merton, P.A., The absence of position sense in
    the human eye. Journal of Physiology, 1960, 153, 127-130.

Brooks, B.A. and Fuchs, A.F., Influence of stimulus parameters on visual sensitivity during saccadic eye movements. Vision Research, 1975, 15, 1389-1398.

Brooks, B.A., Impelman, D.M. and Lum, J.T., Influence of background luminance on visual sensitivity during saccadic eye movements. Experimental Brain Research, 1980, 40, 322-329.

Choe, C.S., Welch, R.B., Gilford, R.M. and Juola, J.F., The "ventriloquist effect": Visual dominance or response bias? Perception and Psychophysics, 1975, 18, 55-60.

Cohen, M.M., Elevator Illusion: Influence of Otolith Organ Activity and Neck Proprioception. Perception and Psychophysics, 1973, 14, 401-406.

Cornsweet, T.N., Determination of the stimuli for involuntary drifts and saccadic eye movements. Journal of the Optical Society of America, 1956, 46, 987-993.

Dell'Osso, L.F., Troost, B.T. and Daroff, R.B., Macro square wave jerks. Neurology, 1975, 25, 975-979.

Ditchburn, R.W., Eye-movements in relation to retinal action. Optica Acta, 1955, 1, 171-176.

Ditchburn, R.W., Eye Movements and Visual Perception. Clarendon Press, Oxford, 1973.

Efron, R., The relationship between the duration of a stimulus and the duration of a perception. Neuropsychologia, 1970a, 8, 37-55.

Efron, R., The minimum duration of a perception. Neuropsychologia, 1970b, 8, 56-63.

Evarts, E.V., Feedback and corollary discharge: A merging of the concepts. Neurosciences Research Progress Bulletin, 1971, 9, 86-112.

Fender, D.H. and Nye, P.W., The effects of retinal image motion in a simple pattern recognition task. Kybernetik, 1962, 1, 192-199.

Festinger, L. and Holtzman, J., Retinal image smear as a source of information about magnitude of eye movement. Journal Experimental Psychology: Human Perception and Performance, 1978, 4, 573-585.

Findlay, J.M., Direction perception and human fixation eye movements. Vision Research, 1974, 14, 703-711.

Fiorentini, A. and Ercoles, A.M., Involuntary eye movements during attempted monocular fixation. Atti Fond. Giorio Ronchi, 1966, 21, 199-217.

Fiorentini, A. and Ercoles, A.M., Visual direction of a point source in the dark. Atti. Fond. G. Ronchi, 1968, 23, 405-428.

Foley, J.M., Primary Distance Perception. In: Handbook of Sensory Physiology. Vol. VIII, Ch. 6, pp. 181-210. R. Held, H.W. Leibowitz and H.L. Teuber (Eds.), Springer-Verlag, Berlin.

Foley, J.M., Successive stero and vernier discrimination as a function of dark interval. Vision Research, 1976, 16, 1269-1273.

Goodwin, G.M., McCloskey, D.I. and Matthews, P.B.C., The contribution of muscle afferents to kinaesthesia shown by vibration induced illusions of movement and by the effects of paralyzing joint afferents. Brain, 1972, 95, 705-748.

Green, D.M., An Introduction to Hearing. Erlbaum Assoc., New Jersey, 1976.

Greenhouse, D.S., Saccadic suppression of flash detection: the uncertainty theory vs. alternative theories. Ph.D. Dissertation, University of California, Berkeley, CA, 1968.

Greenhouse, D.S. and Cohn, T.E., Saccadic suppression: uncertainty vs. alternative theories. April Supp., Investigative Ophthalmology and Visual Science, 1980, 19, 164.

Greenhouse, D.S., Cohn, T.E. and Stark, L., Saccadic suppression may be due entirely to uncertainty of the frame of reference. April Supp., Investigative Ophthalmology and Visual Science, 1977, 17, 106.

Hallett, P.E., Saccades to flashes. In: Eye Movements and Psychological Processes. R.A. Monty and J.W. Senders (Eds.). Wiley, New York, 1976, pp. 255-262.

Hallett, P.E., Primary and secondary saccades to goals defined by instructions. Vision Research, 1978, 18, 1279-1296.

Hallett, P.E. and Lightstone, A.D., Saccadic eye movements towards stimuli triggered by prior saccades. Vision Research, 1976a, 16, 99-106.

Hallett, P.E. and Lightstone, A.D., Saccadic eye movements to flashed targets. Vision Research, 1976b, 16, 107-114.

Hansen, R.M., Spatial localization during pursuit eye movements. Vision Research, 1979, 1213-1221.

Hansen, R.M. and Skavenski, A.A., Accuracy of eye position information for motor control. Vision Research, 1977, 17, 919-926.

Hay, J.C. and Pick, H.L., Jr., Visual and proprioceptive adaptation to optical displacement of the visual stimulus. Journal of Experimental Psychology, 1966, 71, 150-158.

Helmholtz, H. von, Handbuch der Physiologische Optik. Leipzig Voss 1866. English translation from Edit. 3, 1925. Southal, J.P.C. (Ed.): A Treatise on Physiological Optics, 1963, Vol. 3, New York: Dover.

Henson, D.B., Corrective saccades: effects of altering visual feedback. Vision Research, 1978, 18, 63-67.

Heywood, S. and Churcher, J., Direction-specific and position-specific effects upon detection of displacements during saccadic eye movements. Vision Research, 1981, 21, 255-261.

Holly, F., Saccadic presentation of a moving target. Vision Research, 1975, 15, 331-335.

Holst, E. von, Relation between the central nervous system and the peripheral organs. British Journal of Animal Behavior, 1954, 2, 89-94.

Holst, E. von and Mittelstaedt, H., Das Reafferenzprinzip. Naturwissenschaften, 1950, 37, 464-476.

Irvine, S. and Ludvigh, E., Is ocular proprioceptive sense concerned in vision? Arch. Ophthal., 1936, 15, 1037-1049.

Jack, C.E. and Thurlow, W.R., Effects of degree of visual association and angle of displacement on the "ventriloquism" effect. Perceptual and Motor Skills, 1973, 37, 967-979.

Jackson, C.V., Visual factors in auditory localization. Quarterly
    Journal of Experimental Psychology, 1953, 5, 52–65.
James, W., The Principles of Psychology. 1890, Vol. II. Holt. Re-
    printed, New York: Dover 1950.
Jones, B. and Kabanoff, B., Eye movements in auditory space percep-
    tion. Perception and Psychophysics, 1975, 17, 241–245.
Kalil, R.E. and Freedman, S.J., Persistence of ocular rotation
    following compensation for displaced vision. Perceptual and
    Motor Skills, 1966, 22, 135–139.
Keesey, U.T., Effects of involuntary eye movements on visual acuity.
    Journal of the Optical Society of America, 1960, 50, 769–774.
Kinchla, R.A., A Psychophysical Model of Visual-Movement Perception.
    In: Eye Movements and Psychological Processes. R.A. Monty and
    J.W. Senders (Eds.), L. Erlbaum Associates, 1976, pp. 263–275.
Kinchla, R.A. and Allan, L.G., A theory of visual movement percep-
    tion. Psychological Review, 1969, 76, 537–558.
Kinchla, R.A. and Smyzer, F., A diffusion model of perceptual
    memory. Perception and Psychophysics, 1967, 2, 219–229.
Kornheiser, A.S., Adaptation to laterally displaced vision: A
    review. Psychological Bulletin, 1976, 83, 783–816.
Kornmuller, A.E., Eine experimentelle Anesthesie der auberen
    Augenmuskeln am Menschen und ihre Auswirkungen. Journal für
    Psychologie und Neurologie, 1930, 41, 354–366.
Krauskopf, J., Graf, V. and Gaarder, K., Lack of inhibition during
    involuntary saccades. American Journal of Psychology, 1966,
    79, 73–81.
Lamansky, S., Bestimmung der Winkelgeschwindigkeit der Blickbewegung
    respective Augenbewegung. Arch. Ges. Physiol., 1869, 2, 418–422.
Latour, P., Visual threshold during eye movements. Vision Research,
    1962, 2, 261–262.
Ludvigh, E., Possible role of proprioception in the extraocular
    muscles. Archs. Ophthal., 1952, 48, 436–441.
Mack, A., Fendrich, R. and Pleune, J., Adaptation to an altered
    relation between retinal image displacements and saccadic eye
    movements. Vision Research, 1978, 18, 1321–1328.
MacKay, D.M., Theoretical models of space perception. Chapter V,
    pp. 83–103, In: Aspects of the Theory of Artificial Intelli-
    gence. C.A. Muses (Ed.), New York: Plenum Press, 1962, 83–104.
MacKay, D.M., Elevation of the visual threshold by displacement of
    retinal image. Nature, London, 1970, 225, 90–92.
MacKay, D.M., Voluntary Eye Movements as Questions. Karber, Basel,
    Bibl. Ophthal., 1972, 82, 369–376.
MacKay, D.M., Visual stability and voluntary eye movements. In:
    Handbook of Sensory Physiology. R. Jung (Ed.). Berlin:
    Springer-Verlag, 1973, 307–331.
Mateeff, S., Saccadic eye movements and localization of visual
    stimuli. Perception and Psychophysics, 1978, 24, 215–224.
Mateeff, S., Yakimoff, N. and Mitrani, L., Some characteristics of
    the visual masking by moving contours. Vision Research, 1976,
    16, 489–492.

Matin, E., Saccadic suppression: a review and analysis. Psychological Bulletin, 1974, 81, 899-917.

Matin, E., Clymer, B. and Matin, L., Metacontrast and saccadic suppression. Science, 1972, 178, 179-182.

Matin, E., Matin, L., Pola, J. and Kowal, K., The intermittent light illusion and constancy of visual direction during voluntary saccades. Paper presented at Psychonomic Society Meeting, 1969, St. Louis.

Matin, L., Eye movements and perceived visual direction. Ch. 13, pp. 331-380. In: Handbook of Sensory Physiology. D. Jameson and L. Hurvich (Eds.), 1972, Vol. VII/4, Springer-Verlag, Heidelberg.

Matin, L., A Possible Hybrid Mechanism for Modification of Visual Direction Associated with Eye Movements - The Paralyzed Eye Experiment Reconsidered. Perception, 1976a, 5, 233-239.

Matin, L., Saccades and the Extraretinal Signal for Visual Direction. In: Eye Movements and Psychological Processes. R. Monty and J. Senders (Eds.), 1976b, Ch. IV.1, Erlbaum Assoc., New York.

Matin, L., Suppression of the use of extraretinal eye position information (EEPI) for visual localization is normal in normally illuminated visual fields. April Supp. to Investigative Ophthalmology and Visual Science, 1981, 20, 55.

Matin, L. and Bowen, R.W., Measuring the duration of perception. Perception and Psychophysics, 1976, 20, 66-76.

Matin, L. and Kibler, G., Acuity of Visual Perception of Direction in the Dark for Various Positions of the Eye in the Orbit. Perceptual and Motor Skills, 1966, 22, 407-420.

Matin, L. and Matin, E., Visual perception of direction and voluntary saccadic eye movements. In: Cerebral Control of Eye Movements and Motion Perception. J. Dichgans and E. Bizzi (Eds.), 1972, 358-368, Basel: Karger.

Matin, L., Matin, E. and Pearce, D.G., Visual perception of direction when voluntary saccades occur: I. Relation of visual direction of a fixation target extinguished before a saccade to a flash presented during the saccade. Perception and Psychophysics, 1969, 5, 65-80.

Matin, L., Matin, E. and Pearce, D.G., Eye movements in the Dark during the Attempts to Maintain a prior fixation Position. Vision Research, 1970, 10, 837-857.

Matin, L., Matin, E. and Pola, J., Visual perception of direction when voluntary saccades occur: II. Relation of visual direction of a fixation target extinguished before a saccade to a subsequent test flash presented before the saccade. Perception and Psychophysics, 1970, 8, 9-14.

Matin, L. and Pearce, D.G., Visual perception of direction for stimuli flashed during voluntary saccadic eye movement. Science, 1965, 148, 1485-1488.

Matin, L., Matin, E., Pola, J. and Bowen, R., Relative visual direction of two flashes presented at different times or intensities during a voluntary saccade - retinal constraints in the operation of extra-retinal signals. Presented at Eastern Psychological Association Convention.

Matin, L., Pearce, D.G., Matin, E. and Kibler, G., Visual perception of Direction: Roles of Local Sign, Eye Movements and Ocular Proprioception. Vision Research, 1966, 6, 453-469.

Matin, L., Picoult, E., Stevens, J.K., Edwards, M.W., Jr., Young, D. and MacArthur, R., Visual Context Dependent Mislocalizations Under Curare-Induced Partial Paralysis of the Extraocular Muscles. April Supp. to Investigative Ophthalmology and Visual Science, 1980, 19, 81.

Matin, L., Picoult, E., Stevens, J.K., Edwards, M.W., Jr., Young, D. and MacArthur, R., Oculoparalytic illusion: visual-field dependent spatial mislocalization by humans with experimentally paralyzed extraocular muscles. Science, 1981b, in press.

Matin, L., Pola, J. and Matin, E., Changes of Visual Direction with Voluntary Saccadic Eye Movements: Influence of Visual Persistence. Transactions of the American Academy of Optometry, 1972, 49, 897.

Matin, L., Pola, J., Matin, E. and Picoult, E., Vernier discrimination with sequentially-flashed lines: roles of eye movements, retinal offsets and short-term memory. Vision Research, 1981a, 21, 647-656.

Matin, L., Stevens, J.K. and Picoult, E., Perceptual Consequences of Experimental Extraocular Muscle Paralysis. Chapter in: Spatially Oriented Behavior. A. Hein and M. Jeannerod (Eds.), 1982, Springer-Verlag.

McLaughlin, S.C., Parametric adjustment in saccadic eye movements. Perception and Psychophysics, 1967, 2, 359-362.

McLaughlin, S.C. and Webster, R.G., Changes in straight-ahead eye position during adaptation to wedge prisms. Perception and Psychophysics, 1967, 2, 36-44.

Merton, P., The accuracy of directing the eyes and the hand in the dark. Journal of Physiology, 1961, 156, 555-577.

Merton, P., Human position sense and sense of effort. Symposium of the Society for Experimental Biology, 1964, 18, 387-400.

Miller, J.M., Information used by the perceptual and oculomotor systems regarding the amplitude of saccadic and pursuit eye movements. Vision Research, 1980, 20, 59-68.

Mills, A.W., On the minimum audible angle. Journal of the Acoustical Society of America, 1958, 30, 237-246.

Mills, A.W., Lateralization of high-frequency tones. Journal of the Acoustical Society of America, 1960, 32, 132-134.

Mitrani, L., Mateeff, S. and Yakimoff, N., Smearing of the retinal image during voluntary saccadic movements. Vision Research, 1970a, 10, 405-409.

Mitrani, L., Mateeff, S. and Yakimoff, N., Temporal and spatial characteristics of visual suppression during voluntary saccadic eye movement. Vision Research, 1970b, 10, 417-422.

Mitrani, L., Mateeff, S. and Yakimoff, N., Saccadic suppression in the presence of structured background. Vision Research, 1973, 13, 517-521.

Mitrani, L., Yakimoff, N. and Mateeff, S., Dependence of visual suppression on the angular size of voluntary saccadic eye movements. Vision Research, 1970, 10, 411-415.

Monahan, J., Extraretinal feedback and visual localization. Perception and Psychophysics, 1970, 12, 349-353.

Nachmias, J., Determiners of drift of the eye during monocular fixation. Journal of the Optical Society of America, 1961, 51, 761-766.

Nakayama, K. and Balliet, R., Listing's law, eye position sense, and perception of the vertical. Vision Research, 1977, 17, 453-457.

Ogle, K.N., Binocular Vision. Hafner Publishing Company, New York, 1950.

Pearce, D. and Matin, L., Variation of the magnitude of the horizontal-vertical illusion with retinal eccentricity. Perception and Psychophysics, 1969, 6, 241-243.

Pearce, D. and Porter, E., Changes in visual sensitivity associated with voluntary saccades. Psychonomic Science, 1970, 19, 225-227.

Picoult, E., MacArthur, R., Young, D., Edwards, M.W., Jr., Stevens, J.K. and Matin, L., Relation Between Visual and Auditory Maps of Space in Room Illumination and in Darkness. Supplement to Investigative Ophthalmology and Visual Science, 1980, 19, 164.

Platt, B.B. and Warren, D.H., Auditory localization: The importance of eye movements and a textured visual environment. Perception and Psychophysics, 1972, 12, 245-248.

Pola, J., The Relation of Visual Direction to Eye Position During and Following a Voluntary Saccade. Ph.D. Dissertation, Columbia University, 1972.

Pola, J., Voluntary saccades, eye position, and perceived visual direction. In: Eye Movements and Psychological Processes. R.A. Monty and J.W. Senders (Eds.), Wiley, New York, 1976.

Rattle, J.D. and Foley-Fisher, J.A., A relationship between vernier acuity and intersaccadic interval. Optica Acta, 1968, 15, 617-620.

Riggs, L.A., Merton, P.A. and Morton, H.B., Suppression of visual phosphenes during saccadic eye movements. Vision Research, 1974, 14, 997-1011.

Roland, P.E., Sensory feedback to the cerebral cortex during voluntary movement in man. The Behavioral and Brain Sciences, 1978, 1, 129-147.

Salapatek, P., Pattern perception in early infancy. Chapter 3 in: Infant Perception from Sensation to Cognition. L.B. Cohen and P. Salapatek (Eds.), Academic Press, New York, 1975.

Schlodtman, W., Ein Beitrag zur Lehre von der optischen Lokalisation bei Blindgeborenen. Arch.f.Ophth., 1902, 54, 256-267.

Sedgwick, H.A. and Festinger, L., Eye movements, efference and visual perception. In: Eye Movements and Psychological Processes. R.A. Monty and J.W. Senders (Eds.), 1976, pp. 221-230,

Erlbaum Assoc., New Jersey.

Shebilske, W.L., Extraretinal information in corrective saccades and inflow vs. outflow theories of visual direction constancy. Vision Research, 1976, 16, 621-628.

Shebilske, W.L., Visuomotor coordination in visual direction and position constancies. in: Stability and Constancy in Visual Perception. W. Epstein (Ed.). 1977, pp. 23-69, Wiley, New York.

Shebilske, W.L., Sensory feedback during eye movements reconsidered. The Behavioral and Brain Sciences, 1978, 1, 160-161.

Shelton, B.R. and Searle, C.L., The influence of vision on the absolute identification of sound-source position. Perception and Psychophysics, 1980, 28, 589-596.

Sherrington, C.S., Further note on the sensory nerves of the eye muscles. Proceedings of the Royal Society, 1898, 64, 120-121.

Sherrington, C.S., Observations on the sensual role of the proprioceptive nerve supply of the extrinsic ocular muscles. Brain, 1918, 41, 332-343.

Siebeck, R., Wahrnehmungsstorung und Storungswahrnehmung bei Augenmuskellahmungen. Von Graufes Archiv für Ophthalmologie, 1954, 155, 26-34.

Siebeck, R. and Frey, R., Die Wirkungen muskeleschlaffender Mittel auf die Augenmuskeln. Anaesthesist, 1953, 2, 138-141.

Skavenski, A.A., Extraretinal correction and memory for target position. Vision Research, 1971, 11, 743-746.

Skavenski, A.A., Inflow as a source of extraretinal eye position information. Vision Research, 1972, 12, 221-229.

Skavenski, A.A., The nature and role of extraretinal eye-position information in visual localization. Chapter IV.7 in: Eye Movements and Psychological Processes. R.A. Monty and J.W. Senders (Eds.), 1976, Erlabaum Assoc., New Jersey.

Skavenski, A.A., and Steinman, R.M., Control of eye position in the dark. Vision Research, 1970, 10, 193-203.

Skavenski, A.A., Haddad, G. and Steinman, R.M., The extraretinal signal for the visual perception of direction. Perception and Psychophysics, 1972, 11, 287-290.

Stark, L., Kong, R., Schwartz, S., Hendry, D. and Bridgeman, B., Saccadic suppression of image displacement. Vision Research, 1976, 16, 1185-1187.

Stoper, A., Apparent motion of stimuli presented stroboscopically during pursuit movement of the eye. Perception and Psychophysics, 1973, 13, 201-211.

Steinbach, M.J., Pursuing the perceptual rather than the retinal stimulus. Vision Ressearch, 1976, 16, 1371-1376.

Stevens, J.K., Emerson, R.C., Gerstein, G.L., Kallos, T., Neufeld, G.R., Nichols, C.W. and Rosenquist, A.C., Paralysis of the awake human: Visual perceptions. Vision Research, 1976, 16, 93-98.

Stevens, J.K., The corollary discharge: is it a sense of position or a sense of space? The Behavioral and Brain Sciences, 1978, 1, 163-165.

Taylor, J.G., The Behavioral Basis of Perception. Yale University
    Press, 1962, New Haven.
Thurlow, W.R. and Jack, C.E., Certain determinants of the "ventrilo-
    quism effect". Perceptual and Motor Skills, 1973, 36, 1171-1184.
Uttal, W. and Smith, P., Recognition of alphabetic characters during
    voluntary eye movements. Perception and Psychophysics, 1968,
    3, 257-264.
Volkmann, F., Vision during voluntary saccadic eye movements.
    Journal of the Optical Society of America, 1962, 52, 571-578.
Volkmann, F., Schick, A.M.L. and Riggs, L.A., Time course of visual
    inhibition during voluntary saccades. Journal of the Optical
    Society of America, 1968, 58, 562-569.
Wallach, H. and Lewis, C., The effect of abnormal displacement of
    the retinal image during eye movements. Perception and Psycho-
    physics, 1966, 1, 25-29.
Ward, F., Pursuit eye movements and visual localization. In: Eye
    Movements and Psychological Processes. R.A. Monty and
    J.W. Senders (Eds.). Erlbaum Assoc., New Jersey, 1976, pp.
    289-297.
Weber, R.B. and Daroff, R.B., The metrics of horizontal saccadic eye
    movements in normal humans. Vision Research, 1971, 11, 921-928.
Weber, R.B. and Daroff, R.B., Corrective movements following refixa-
    tion saccades: type and control system analysis. Vision
    Research, 1972, 12, 467-475.
Welch, R.B., Perceptual modification: Adapting to altered sensory
    environments. Academic Press, 1978, New York.
Wertheim, A.H., On the relativity of perceived motion. Acta
    Psychologica, 1981, 48, (Special Issue on the Perception of
    Motion) 97-110.
Wertheimer, M., Psychomotor coordination of auditory and visual
    space at birth. Science, 1961, 134, 1692.
Whipple, W.R. and Wallach, H., Direction-specific motion thresholds
    for abnormal image shifts during saccadic eye movement.
    Perception and Psychophysics, 1978, 24, 349-355.
Witkin, H.A., Perception of body position and of the position of
    the visual field. Psychological Monographs, 1949, 63, no. 7.
Witkin, H.A. and Asch, S.E., Studies in space orientation. IV.
    Further experiments on perception of the upright with dis-
    placed visual fields. Journal of Experimental Psychology, 1948,
    38, 762-782.
Witkin, H.A., Wapner, S. and Levinthal, T., Sound Localization with
    conflicting visual and auditory cues. Journal of Experimental
    Psychology, 1952, 43, 58-67.
Wolf, W., Hauske, G. and Lupp, U., How presaccadic gratings modify
    postsaccadic modulation transfer function. Vision Research,
    1978, 18, 1173-1180.
Yakimoff, N., Mitrani, L. and Mateeff, S., Saccadic suppression as
    visual masking effect. Agressologie, 1974, 15, 387-394.
Zahn, J.R., Abel, L.A. and Dell'Osso, L.F., Audio-ocular response
    characteristics. Sensory Processes, 1978, 2, 32-37.

Zuber, B. and Stark, L., Saccadic suppression: Elevation of visual
    threshold associated with saccadic eye movements. Experimental
    Neurology, 1966, 16, 65-79.

LINEAR SELF MOTION PERCEPTION

Alain Berthoz and Jaques Droulez

Laboratoire de Physiologie Neurosensorielle du CNRS

15 rue de l'Ecole de Médecine, 75006 Paris, France

INTRODUCTION

The purpose of this paper is to review some aspects of the current knowledge concerning linear self motion perception. An adequate perception of self motion is important for movement, locomotion (Review in Schöne, 1980), and in any situation in which man has to drive a machine or a vehicle; it requires the evaluation of head motion in space which can be accomplished either by the visual or the vestibular systems. The case of angular rotation has been extensively investigated: the main features of the semi-circular canals, their dynamic properties and the effect of their stimulation (by angular acceleration) on perception or motor control are well known (see for some reviews: Baker and Berthoz, 1977; Wilson and Melvill Jones, 1979). The characteristics of visually induced circular self motion perception (circular vection), and the interaction between semi-circular canals and vision in self motion perception have been reviewed by Dichgans and Brandt (1978) and by Leibowitz et al. (this volume).

In contrast with circular motion, the perception of linear motion is not well documented. Only partial data are available concerning the properties of the otolithic system: the utriculus and sacculus have been, for a long time, considered only as gravitational "static" receptors, and the related perceptions were studied mainly in connection with apparent vertical estimation (see reviews by Guedry, 1974 and Graybiel, 1974). Not only do we know little about the dynamic properties of the otolithic system but we also lack information concerning the central treatment of the sensory information on linear motion provided by the otoliths. Although it is now well established that their specific stimulus is the shear force induced by a linear acceleration acting in the plane of their maculae, a basic ambiguity (Barlow, 1964) is built in with this type of sensor acting in the gravitational field: the generalized relativity theory (Einstein, 1945) predicts that each receptor cannot distinguish between a true linear acceleration and a component of gravity which acts in the plane of the macula during a static head tilt (Fig. 1).

In spite of this ambiguity, human subjects can differentiate these two situations perceptually: during dynamic otolithic stimulation, e.g. changing the overall linear acceleration in magnitude or duration, the subject experiences little or no tilt but rather linear motion (Von Bekesy, 1940; Lansberg, 1954; Guedry and Harris, 1963). By contrast, in centrifuge experiments, subjects, submitted to a constant horizontal acceleration which is added to gravity, experience no linear motion but rather modifications of the apparent vertical related to the resultant acceleration vector (sum of gravity and linear acceleration vector). It should be stressed that on earth no pure linear acceleration stimulus can be given. Changing linear acceleration induces a "rotating vector" by its combination with gravity. Even during vertical acceleration the stimulus is biased by a constant colinear acceleration. One may therefore conclude that complex interactions between the vestibular receptors themselves probably play a major role in the separation of these situations (Mayne, 1974; Young, 1974). Moreover, the additional role of vision in this process is practically unknown. Although the "optic flow" which is the visual input during a linear displacement has been analysed by several authors (Gibson, 1957; Gibson and Gibson, 1966; Hay, 1966; Lee, 1974; Longuet-Higgins and Pradzny, 1980) only a few experimental studies have been conducted on humans and/or animals.

In addition, both vestibularly and visually induced linear motion perception are accompanied by postural readjustments (Helmholtz, 1896; Fisher and Kornmüller, 1930; Lishman and Lee, 1973; Lestienne et al., 1977) and oculomotor effects (Niven et al., 1966; Buizza et al., 1980). These motors effects will not be reviewed here but only mentioned for some particular theoretical considerations (see for a review: Berthoz et al., 1979).

1. ROLE OF THE OTOLITHS IN LINEAR SELF MOTION PERCEPTION

1.1. Specific stimuli for the otolith receptors

After a century of discussions it is now accepted that the specific stimulus for the activation of the otolith end organs is the shear force between the otoliths and the maculae. This force is induced by the components of linear acceleration (including gravity) acting in the plane of the maculae (Fig. 1).

The utriculus and the sacculus are complex curved surfaces. One can assume roughly that the utriculus is in the head horizontal plane when the head is tilted about 30° forward and that the sacculus is in the head vertical plane parallel to the anterior semicircular

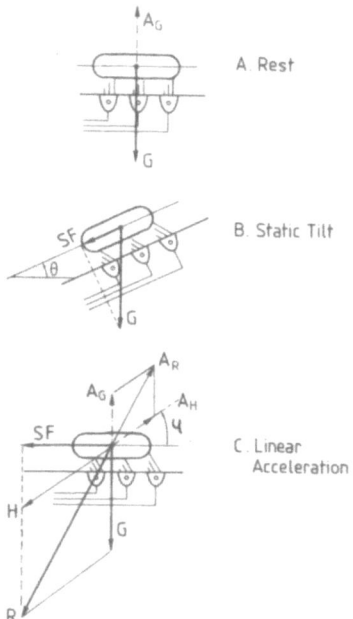

Fig. 1. Combination of the force vectors of gravity and linear acceleration to produce the specific shear force stimulus on the
otolith end organs.
A: Head upright, without motion. G is the weight of the
otolith acting perpendicularly to the macular plane; A_G is
the inertial head acceleration which would produce the same
force G.
B: Tilted head, without motion. SF (the Shear Force acting
on the ciliated cells) is equal to the projection of G on
the macular plane.
C: Head upright, with motion. R (the overall resulting
force acting on the macular cells) is equal to the vectorial
sum of gravity force (G) and inertial force due to head
motion (H). A_R, A_G, A_H are the corresponding acceleration
vectors. SF, the shear force, is equal to the projection of
R on the macular plane.

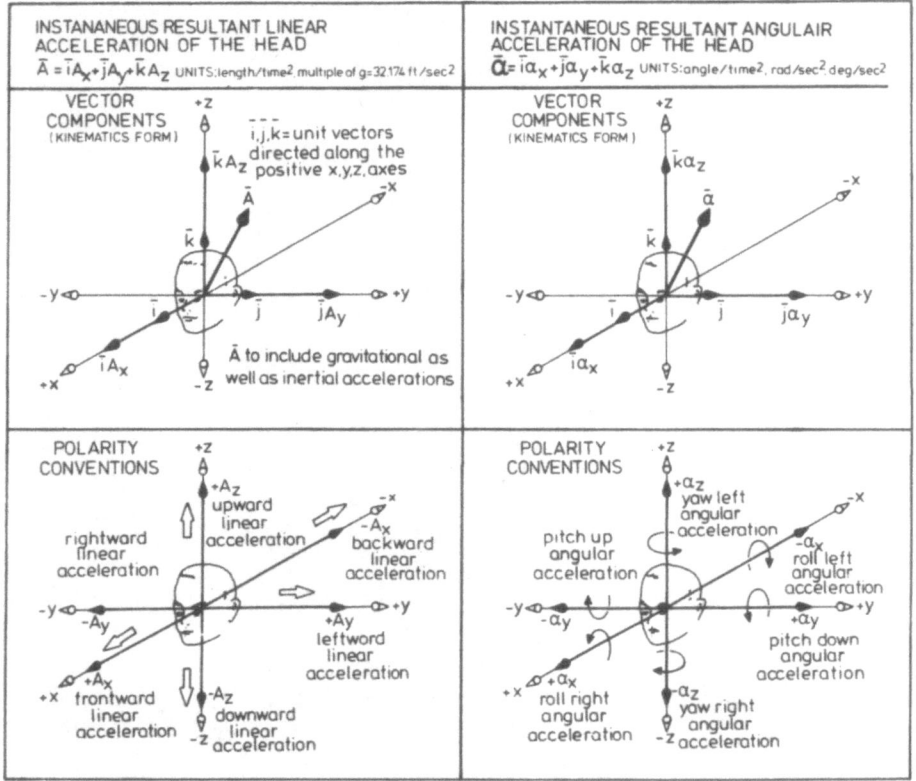

Fig. 2. Vestibular nomenclature indicating the stimuli notation for
linear and angular acceleration (from Hixson et al., 1966).

canal, that is, at 45° from the sagittal plane. Conventional head
motion notation (Hixson et al., 1966) and otolith planes are shown
in Fig. 2.

The main current knowledge concerning these receptors has been
reviewed in Wilson and Melvill Jones (1979). We cannot attribute to

each otolith organ a unique direction of sensitivity. Each receptor
can be characterized by "functional sensitivity vectors", also called
"functional polarisation vectors" which characterize individual hair
cells. These vectors can be drawn in polar coordinates in the plane
of each macula (as a first approximation). The length of the vector
gives the change in firing rate of afferent VIIIth nerve fibers per
change of the component in linear acceleration directed along the
line. The angle or phase of the vector indicates the direction in
which sensitivity is maximal. With this representation, Fernandez
and Goldberg (1976a) have shown in the monkey that functional polarisation vectors for utriculus and sacculus are generally in an
orthogonal arrangement corresponding to the planes of the maculae,
and that both organs are probably involved in a similar acceleration
detection process each with its own geometrical features.

At the level of the vestibular nuclei the problem is more complex because obviously a high degree of processing occurs within
the vestibular nuclei and output information from otolith sensitive
units is different in its dynamic properties from afferent information. For instance there seems to be two types of cells: "dynamic"
and "static" which may be related to the necessary segregation of
gravito-inertial and accelerative stimuli by the central nervous
system as discussed above. As the purpose of this paper is not to
review neurophysiological data in detail, the reader is referred to
the review by Wilson and Melvill Jones (1979). It should be enough
to state that the available data concerning these receptors validate
the low perception thresholds which will be described below.

1.2. Non-otolithic contribution to self motion perception in darkness

Before describing the various threshold measurements which have
been performed concerning otolithic detection of linear motion it
may be useful to review possible non-otolithic contributions to
linear motion perception. Three main sources of sensory information
will be considered: the semi-circular canals and the somatosensory
and auditory systems.

1.2.1. Semi-circular canals

Additional information acquired during linear motion complicates
the detection of motion by the otolithic organs and may obliterate
their contribution during psychophysical experiments. In the absence of rotation it is possible that the semi-circular canals contribute to a "non-rotation" signal. In addition, Steer (1967) has
proposed a dynamic semi-circular canal contribution in linear horizontal motion. The mechanism would be a displacement of the endo-
lymph generated by the rotating resultant of linear acceleration and
gravity. Similar phenomena are supposed to be involved in "barbecue"
rotation (Benson, 1974) ("barbecue" rotation is when the subject is
rotated around his body axis but perpendicular to gravity). However,
such a semi-circular canal contribution cannot account for vertical
linear motion perception because in this latter case there is no

rotating linear acceleration vector. Recent recordings of the vestibular primary afferent response in the monkey (Goldberg and Fernandez,
1981) lead to the conclusion that the contribution of the canals to
the detection of linear acceleration is minor or artefactual.

1.2.2. Somatosensory system

During a linear acceleration the somato-sensory system and
particularly the cutaneous receptors may be stimulated and contribute to the perception of linear motion. Attempts have been made to
investigate the possible somato-sensory contribution to the thresholds for linear motion detection. The experiments were done with
pathological subjects with high spinal lesions. These subjects exhibit linear acceleration detection thresholds which are very similar
(4 to 8 cm/sec²) to those of normal subjects (2 to 5 cm/sec²)
(Walsh, 1961). In the same experiments, Walsh tried to immerse subjects in water or to wrap them in cotton in order to modify the
pressure distribution on the skin. This did not change the results
significantly.

Another way to demonstrate the otolithic origin of linear motion
detection thresholds is to reduce the otolithic information. This
has been done in labyrinthine defective subjects which were shown to
have thresholds two orders of magnitude higher (200 cm/sec²) than
normal subjects (Guedry and Harris, 1963). Guedry (1974) and Jongkees
and Groen (1950) reported that labyrinthine defective subjects experienced little motion perception but rather head tilt. Perceptual
thresholds are also dependent upon the posture of the subjects.
Standing subjects in darkness have lower thresholds than when seated
in the same experimental condition (see Table I). This may obviously
be due to the stimulation of the somatosensory system (foot pressure
receptors, muscle and joint proprioception) (Brandt et al., 1977).
Another possible cue for linear motion detection is the information
called "haptokinetic" (Gibson, 1954; Berthoz, 1978), namely the
rubbing of textured surfaces on the skin during linear motion.

1.2.3. Auditory cues

The contribution of auditory cues to self motion perception is
yet unclear. It is known that "audio-kinetic" circular self motion
perception can be elicited by a rotatory acoustical stimulation
(Urbantschitsch, 1897; Von Stein, 1910; Dodge, 1923). However, no
information is available concerning linear motion. The fact that
moving acoustic targets induce mainly saccadic eye movements
(Buizza et al., 1979) may mean that the reconstruction of a self
motion velocity vector based on auditory cues is not an easy task
for the central nervous system. Some differences may exist between
localized acoustic targets and full sound field motion.

1.3. Threshold measurements

Guedry (1974) and Gundry (1978) reviewed experimental measurements of linear motion perception thresholds. The results obtained

Table I. Thresholds of perception of linear oscillations in man in the absence of visual information. (From Guedry (1974) with additional references.)

Authors	Year	Number of subjects	Subject posture	Gravity orientation (subject relative)	Oscillations orientation (subject relative)	Oscillations orientation (gravity relative)	Frequency (Hz)	Period (s)	Acceleration threshold of motion perception (cm/sec²)
MACH	1875	2		Z	Z	vertical	0.14	7	10-12
GURNEE	1934	3 3	Seated Seated	Z Z	Z Z	vertical vertical	0.125 0.062	8 16	8-10 6-7
TRAVIS and DODGE	1928	2 2 2	Seated Standing Standing	Z Z Z	X X Y	horizontal horizontal horizontal	0.5-0.125 0.5-0.125 0.5-0.125	2-8 2-8 2-8	20-25 8 5
JONKEES and GROEN	1946	2		X	Z	horizontal	0.4	2.5	6-13
LANSBERG	1954		Laying	±X	Z	horizontal	0.26	3.8	9-15
WALSH	1961	4-7	Laying Laying Laying	±X or ±Y ±X ±Y	Z Y X	horizontal horizontal horizontal	0.4 0.4 0.4	2.5 2.5 2.5	2.2 2.0 1.8
WALSH	1962	6 6	Laying Laying	±X or ±Y ±X or ±Y	Z Z	horizontal horizontal	0.33 0.11	3.0 9.0	7-8 9
WALSH	1964	7	Laying	X	C	vertical	0.4 0.33 0.11	2.5 3.0 9.0	6 7 18
BENSON, DIAZ and FARRUGIA	1975	12	Seated	Tilted from Z	X or Y	Stimulation produced by off-vertical axis rotation	$5.5 \cdot 10^{-3}$ $2-8 \cdot 10^{-2}$ 0.125 0.83	180 36 8 1.2	100-120 70 30 12
HOSMAN and VAN DER VAART	1978	3	Seated	Z	Z	vertical	0.16 0.6 2.0	6.3 1.6 0.5	3-7 5-9 3-9

in darkness or with blindfolded subjects are summarized in Table I. On first inspection these data show a wide variability. As examples of the numerous factors involved, the timing and the direction of stimulation are discussed separately below. The differences in the methods used to stimulate the otolith organs (linear moving carts, parallel swings, elevators, centrifuges, helicopters, planes and other stimulating devices) could account for part of this variability. Note the trend for lower thresholds in more recent studies, which may be related to technical improvements.

The other set of variability factors are inherent to all perception studies: intersubject variability, influence of the instructions, attention and superimposed tasks, methods of recording the perceptual responses, etc. Concerning the last point, it is very important to note that a great difference may exist between the thresholds for detection of *motion* and those for an adequate detection of *motion direction*. These two types of thresholds are often distinct and have to be measured spearately (Malcolm and Melvill Jones, 1974; Jones and Young, 1976).

In conclusion, the otolithic system seems to be the most sensitive and specific system for linear motion perception in darkness. However, one should keep in mind that in normal behaviour linear motion detection involves multi-sensory interactions and that other motion cues including the subjects' conceptual knowledge about his own motion may be implied in this process (Guedry, 1974).

In addition, we deal here only with *passive* linear motion perception while linear motion is generally the result of an *active* motor action like locomotion.

1.4. Interaxis differences

In spite of the wide variability mentioned above it is possible to compare the results obtained by the same authors using the same stimuli applied to the same subjects but in different directions with respect to the subjects head, or with respect to gravity. Walsh (1962) reported a slightly higher threshold when horizontal linear acceleration was applied along a Z axis while gravity was in X or Y direction relative to the subject's head (see Table I). A similar result was found by Meiry (1965) during sustained linear acceleration, but this difference is not very significant compared to the variability in threshold data.

As mentioned above, Malcolm and Melvill Jones (1974) showed that upright seated subjects submitted to vertical acceleration along a Z axis (up to 4 m/sec²) were not able to report accurately their direction of motion. Like Walsh (1962), Meiry (1965) did not observe this handicap when subjects lay supine and were submitted to horizontal acceleration along a Z axis. It has been suggested that this effect is due to the orientation of the head with respect to gravity rather than to the orientation of the applied acceleration with respect to the head (Jones and Young, 1976). The physical input to the otolith differs in the two experimental conditions: in the supine position the otoliths are stimulated by a rotation vector, while in the vertical normal position the direction of the stimulus is invariant and is colinear to gravity with only a constant bias.

1.5. Dynamics of vestibular linear motion perception

Short term dynamics of vestibular linear motion perception have been investigated during threshold and supra-threshold experiments. Long term dynamics (adaptation, habituation) studies give rise to technical difficulties. At frequencies below 0.05 Hz for example, the data from Benson et al. (1975) in centrifuge experiments are the only ones available. The development of human space flight gives the scientist the opportunity to investigate long term adaptation to weightlessness.

1.5.1. Dynamics inferred from threshold experiments

Motion detection thresholds have been measured as a function of *linear oscillation* frequency. In spite of the variability in the results, self motion detection thresholds seem to decrease when the stimulation frequency increases up to 1 Hz. Data from Benson et al. (1975) and Walsh (1961) yield to a roughly linear relationship between threshold and frequency with a slope about 0.5. This suggests that what is actually detected is a combination of jerk and acceleration (Gundry, 1978). Above 1 Hz reliable data are lacking (most experiments have been performed in "stabilised vision" i.e. with a subject standing in a closed illuminated room moving with him).

During *sustained linear acceleration* the latency of onset of self motion perception has been plotted as a function of the magnitude of the acceleration level (see Fig. 3). Data from Meiry (1965) and Jones and Young (1976) lead to a linear relationship between acceleration and latency expressed in log coordinates. This suggests a velocity threshold mechanism. It is striking that similar results have also been found in circular motion detection (Meiry, 1965; Ormsby, 1974).

A comparison of the two sets of threshold data (e.g. linear oscillation and sustained linear acceleration) reveals some discrepancy in the results: the first set shows a relationship between the threshold for the perception of motion and peak acceleration, the latter set shows a velocity threshold mechanism. However, one must keep in mind two facts. First, during linear oscillation experiments with a constant frequency, the stimulus is predictable. As pointed out by Guedry (1974), stimulus predictability lowers the threshold and may change the dynamics of self motion detection. Second, during sustained linear acceleration the stimulus is an acceleration step. Thus the subjects are submitted to an uncontrolled jerk which is dependent upon the technical set up and the subject restraining equipment. Consequently the relationship between jerk and acceleration may be masked in these experiments.

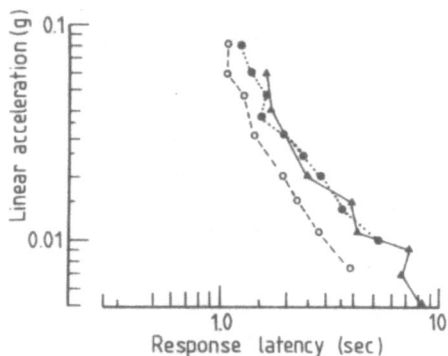

Fig. 3. *Perception of linear acceleration in man. Response latency for the detection of step inputs of sustained linear acceleration.* Response latency and threshold of acceleration detection for stimuli applied along X (1) and Z (2) body axis. Data from Meiry (1965): X (1), solid circles; Z (2), open circles, with 3 subjects. From Melvill Jones and Young (1976): same measurements from 8 subjects seated with head tilted forward by 30° (to bring the utricular macula into the earth horizontal plane) submitted to acceleration along the Z body axis (triangles). A regression line fitted through all points indicated a threshold which had the dimensions of a velocity of .21 m/sec coupled with a delay interpreted by the authors as a minimum reaction time of 0.375 sec.

1.5.2. Dynamics inferred from suprathreshold experiments

The phase relationship of perceived self motion as compared with the subject's actual motion has been investigated by Meiry (1965) (see experimental data in Fig. 4). Upright seated subjects were submitted to horizontal linear oscillations along the X axis. Due to the limited length of the track, it increased only from 0.1 to 2 m/sec² between 0.025 and 0.1 Hz. Similar experiments were performed by Young, Meiry and Li (1966) by applying linear earth horizontal accelerations along the Z axis of supine subjects. When earth vertical Z axis accelerations were applied to upright seated subjects

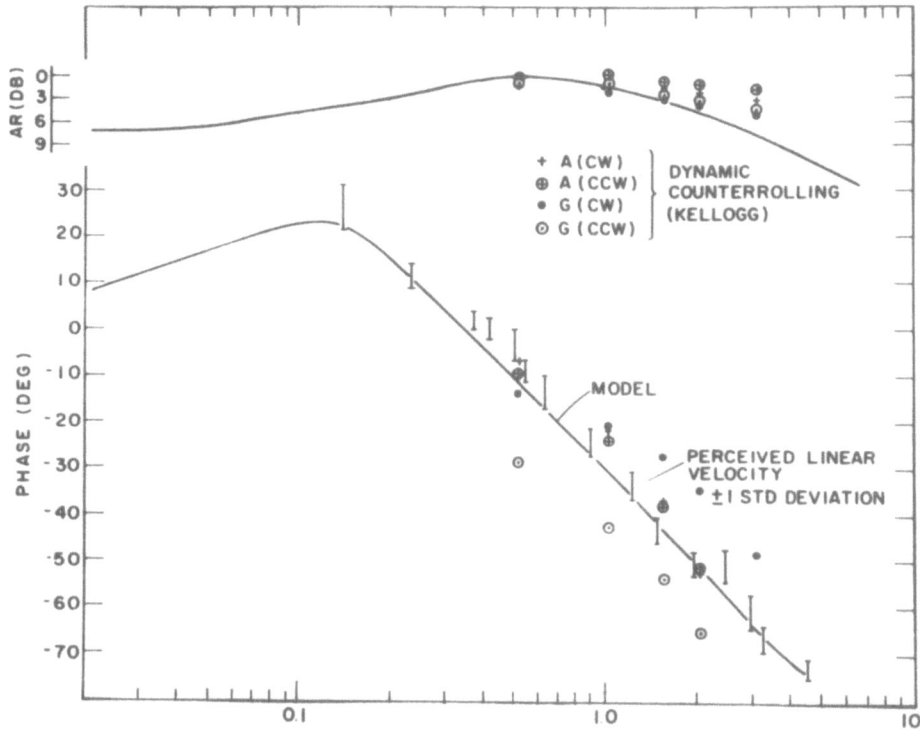

Fig. 4. *Phase response of the detection of linear acceleration in man.* Comparison of experimental results on perceived linear velocity from Meiry (1965) and on dynamic counter-rolling from Kellogg (1965) with the revised dynamic otolithic model (Young and Meiry, 1968). CW and CCW refer to clockwise and counter-clockwise stimulations in dynamic counter-rolling data (2 subjects).

(Melvill Jones et al., 1980) the phase relationship between perceived and actual self motion was very variable. Such discrepancies, which have already been mentioned above when discussing threshold experiments, could mean that linear motion perception may be related to the evaluation by the central nervous system of *changes in the direction of the resultant vector* (linear acceleration plus gravity)

rather than the magnitude of the *linear acceleration vector* itself
as separated from the gravitational component. Guedry (1974) points
out that there are also some discrepancies between Meiry's (1965)
data and Wash's (1962) experiments.There appears to be a good fitting
of Walsh's data with those of Meiry, when the time of detection of
motion reversal during sinusoidal oscillation is plotted against
frequency. However, at 0.11 Hz, Walsh reports a 180° phase lag (or
lead) between perceived and actual motion, instead of about 20° as
exhibited by Meiry's data in Figure 4.

2. MODELS OF LINEAR MOTION PERCEPTION OF OTOLITHIC ORIGIN

The elaboration of models of vestibular psychophysics has proven
to be successful for rotations in an earth horizontal plane about
the subject's X, Y or Z axis. Transfer functions of circular self
motion perception induced by stimulation of the semicircular canals
have been described, in which contributions of end organ (canal)
dynamics as well as central processing have been formalised separately (see Meiry, 1965).

Models of linear motion detection do not reach this point of
achievement, although they follow the same approach. Some difficulties arise in the attempt to formalise end organ dynamics and
central processing when neurophysiological data like those of
Fernandez and Goldberg (1976) are taken into account. Another
problem concerns the intravestibular interactions which generally
occur in most cases of otolithic stimulations. To our knowledge,
this problem has only been handled by Ormsby (1974), and Mayne
(1974).

We have tried to clarify the problems of the otolith psychophysics models by considering separately: a) Afferent processing
(transfer function of the mechanical end organ and transfer function
of the transducer), and b) Central processing (further central
treatment of the otolith signal and intravestibular interactions).
A comparison with neurophysiological data will be briefly presented.
Table II summarizes the transfer functions and the time constants
as proposed by different authors.

2.1. Afferent processing

2.1.1. End organ mechanical system

The mathematical description of otoconia displacement is (in a
formal sense) similar to that of cupula displacement, e.g. the otolith end organ is assumed to be a viscous damped inertial linear
accelerometer (Meiry, 1965; Ormsby, 1974).

Hence the transfer function of the mechanical system takes the
general form of the "torsion pendulum" model (k is a constant)

$$\frac{x(s)}{a(s)} = \frac{kT_1T_2}{(1 + sT_1)(1 + sT_2)}$$

Table II. Transfer function and time constants of otolith models.

AUTHORS	YEAR	EXPERIMENTAL DATA	TRANSFER FUNCTION	NUMERICAL VALUES
MEIRY	1965	Linear motion perception	$\dfrac{Ks}{(1+sT_1)(1+sT_2)}$	$T_1 = 10$ sec $T_2 = 0.66$ sec
MAYNE and BELANGER	1966	Neurophysiological data from Löwenstein and Roberts (1950)	Equivalent to: Type 2: $\dfrac{K}{1+sT_1}$ Type 3: $\dfrac{K(1+sT_2)}{(1+sT_1)}$	$T_1 = 4$ sec $T_1 = 4.3$ sec $T_2 = 7.8$ sec
YOUNG and MEIRY	1968	Linear motion perception from Meiry (1966) counterrolling from Kellog (1865)	$\dfrac{K(1+sT_3)}{(1+sT_1)(1+sT_2)}$	$T_1 = 5.3$ sec $T_2 = 0.66$ sec $T_3 = 13.2$ sec
ORMSBY	1974	Various illusions and motion perception	Equivalent to: $\dfrac{K(1+sT_3)}{(1+sT_1)(1+sT_2)}$	$T_1 = 7.5$ sec $T_2 = 0.51$ sec $T_3 = 10.1$ sec
FERNANDEZ and GOLDBERG	1976	Neurophysiological data (primary afferent fibers)	$\dfrac{K(1+k_a T_a s)(1+\mu(T_2 s)^{\mu})}{(1+T_a s)(1+T_1 s)}$	Regular units $T_1 = 0.016$ sec $T_a = 69$ sec $T_2^* = 40$ sec $\mu = 0.188$ $K_a T_a = 77$ sec Irregular units $T_1 = 0.009$ sec $T_a = 101$ sec $T_2^* = 40$ sec $\mu = 0.44$ $K_a T_a = 192$ sec

* = Arbitrarily chosen between 5 and 320

where x(s) and a(s) are the Laplace transforms of otolith displacement (skull relative) and head linear acceleration (or more precisely
the projection of head linear acceleration on the plane of the utricular or saccular epithelioma). The longer time constant (T_2) has
been estimated by various techniques which gave values ranging
between 3 and 10 msec. In other words, within the frequency range of
natural movements, the short time constant (T_1) can be neglected.
The mechanical model can thus be approximated by a first order linear
system:

$$\frac{x(s)}{a(s)} = \frac{k(T_m)}{1 + sT_m}$$

2.1.2. Transducer and afferent processing

The first model proposed by Meiry (1965) was built in order to
simulate linear self motion perception, and no attempt was made by
this author to include steady state (tilt or centrifuge experiments) or neurophysiological data. In their revised model, Young and
Meiry (1968) included a "lead term": $(1 + sT_3)$ which may be related
to the transducer dynamics. However, they favor a central processing
explanation, based upon threshold considerations; hence this lead
term is discussed below.

Macula hair are known to behave as kinetic (phasic) and position (tonic) transducers (Löwenstein and Roberts, 1949). In the cat,
Vidal et al. (1971) found a simple linear relationship between firing
rate ($F(t)$) of afferent fibers and the tilt angle $\theta(t)$:

$$F(t) = G(\theta(t) + K \frac{d\theta}{dt} (t))$$

with an nondirectional sensitivity to the angular velocity $\left(\frac{d\theta}{dt}\right)$.
This relationship can be written with the Laplace transformation
(G is a constant):

$$\frac{F(s)}{\theta(s)} = G(1 + Ks)$$

A similar relationship was found in the bullfrog by Löwenstein and
Saunders (1975). Fernandez and Goldberg (1976) recorded the discharge rate of primary afferent fibers from otolith nerves in the
squirrel monkey (see Fig. 5). They found two main types of units:
-    regular, tonic, weakly adapting units,
-    irregular, phasic and adapting units.

The relationship between the discharge rate of those two types
of units to the applied specific force may be expressed in terms of
a general transfer function:

$$H(s) = K \cdot \frac{(1 + K_a T_a s)}{(1 + T_a s)} \cdot \frac{(1 + \mu (T_2 s)^{\mu})}{(1 + T_1 s)}$$

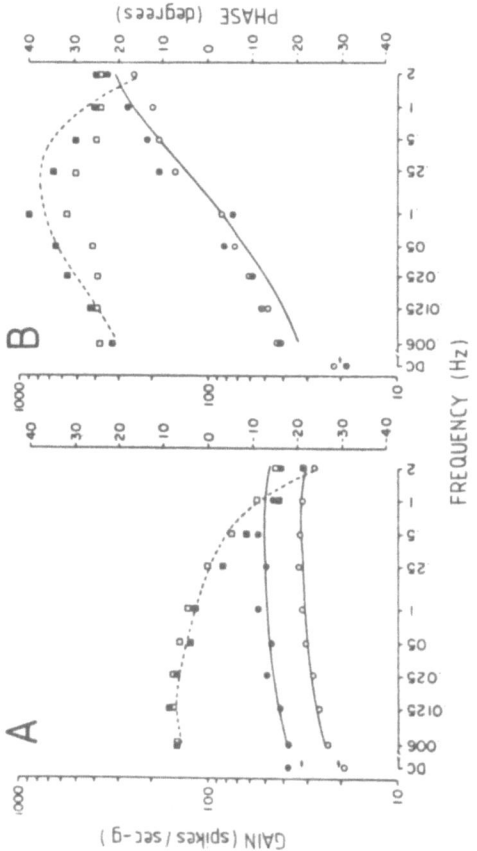

Fig. 5. Bode plots of peripheral otolith neurons in the monkey. A: regular unit. B: irregular unit. Solid and open symbols, respectively, excitatory and inhibitory sine waves. Excitatory and inhibitory sine waves refer to the discharge pattern of units when the applied centrifugal force was respectively parallel and antiparallel to the unit's polarization vector. Curves derived from transfer function (see Table II). Arrows: assumed DC gain. (From Fernandez and Goldberg, 1976.)

Values of the gains (K, K_a), time constants (T_1, T_2, T_a) and non linearity terms (μ) are given in Table II for regular and irregular units. A similar classification has been made for otolith primary afferent fiber responses in the cat (Anderson et al., 1978).

The general transfer function mentioned above can be decomposed into:

- an adaptive term: $\dfrac{1 + K_a T_a s}{1 + T_a s}$

- a non-linear phasic term: $1 + \mu(T_2 s)^\mu$

- a damping factor: $\dfrac{1}{1 + T_1 s}$ with a short time constant (about 10 msec)

which may be related to the dynamics of the mechanical displacement of the otolith.
Note that T_2 was arbitrarily chosen by Fernandez and Goldberg (see Table II) and can be varied from 5 to 320 sec with equally good fits.

The non-linear phasic term: $1 + \mu(T_2 s)^\mu$ could correspond to a "lead term"; however, it is not evident that this one corresponds to the "lead term" proposed by Young and Meiry (1968).

Mayne and Belanger (1966) proposed the following transfer function, primarily based on recordings from fish preparations by Löwenstein and Roberts (1949), to express the response of otolithic cells to their specific stimulus (sum of gravity and acceleration):

$$\frac{x(s)}{a(s)} = K_1 - \frac{K_2 \omega_1 \omega_2}{(s + \omega_1)(s + \omega_2)}$$

Depending on the numerical values of the constants $(K_1, K_2, \omega_1, \omega_2)$, such a transfer function can be applied to three types of cell responses:

type 1: *static (or tonic) cells: $K_2 = 0$*
      The cell response is proportional to the resultant acceleration (sum of gravity and head acceleration).

type 2: *dynamic (or phasic) cells: $K_1 = K_2 = K$*
      The transfer function can be rewritten as follows:

$$\frac{x(s)}{a(s)} = K \cdot \frac{s}{1 + T s}$$

The cell response corresponds to the rate of change of the resultant acceleration, with a time constant T of about 4 sec.

type 3: *mixed (tonic-phasic) cells:* $K_1 < K_2$
Such cells respond to both acceleration and rate of change of acceleration.

As mentioned by Mayne (1974), the differentiation of cell responses into these three types can be recognized as a first attempt to sort out the gravitational and inertial components of the otolithic specific stimulus. However, during head inclination, the gravity component acting along a given axis is not a constant. Therefore, the gravito-inertial separation problem requires further processing. This point will be discussed later.

2.2. Central processing

2.2.1. Central processing of otolith signals

As in the case of circular motion perception, Meiry (1965) proposed to correlate perceived linear velocity with otoconia displacement. Thus, the transfer function of perceived linear velocity takes the form:

$$\frac{\hat{V}(s)}{V(s)} = K \frac{x(s)}{v(s)} = \frac{Ks}{(1 + sT_1)(1 + sT_2)}$$

where $\hat{V}(s)$ and $V(s)$ are the Laplace transforms of perceived head velocity and actual head velocity. Evidently such a model cannot predict the steady state perception of tilts (τ_x and τ_y) according to the terminology of Guedry (1974).

The lead term introduced later by Young and Meiry (1968) led to the following transfer function:

$$\frac{\hat{V}(s)}{x(s)} = \frac{K(1 + sT_3)}{s}$$

or

$$\frac{\hat{V}(s)}{V(s)} = K \cdot \frac{1 + sT_3}{(1 + sT_1)(1 + sT_2)}$$

As indicated in Table II, the value for T_3 (13.2 s) involves a somewhat similar behavior of the first model (Meiry, 1965) and the revised model (Young and Meiry (1968)) in the high frequency range (> 0.2 Hz). However, in the steady state or low frequency range, the revised model can predict perceived body tilt ($\hat{\theta}$) with the following transfer function:

$$\frac{\hat{\theta}(s)}{\theta(s)} = \frac{\hat{a}(s)}{a(s)} = \frac{K(1 + sT_3)}{(1 + sT_1)(1 + sT_2)}$$

Furthermore, Young and Meiry (1968) introduced a threshold before the lead term, in order to account for the non-linearity in

the perception of linear body motion. However, it must be stressed
that the time constants of their revised model do not correspond to
those computed by Fernandez and Goldberg (1976) from primary afferent fiber responses.

In cats submitted to surgical blockage of the semicircular
canals (method of Money and Scott, 1962), Schor (1974) recorded
vestibular units during dynamic tilts of the head. As in primary
afferent fibers, a dynamic type and a static type response were
found. Schor observed that when cell responses are plotted against
the frequency of stimulation in a log-log plot, the dynamic index,
given by the slope of this plot, is much higher in vestibular nuclei
cells than in primary afferent fibers.

The responses of dynamic-type cells were also recorded in
vestibular nuclei of cats submitted to pure linear acceleration
(Adrian, 1943; Melvill Jones and Milsum, 1969; Daunton and Thomsen,
1976). However, in such conditions, a strong phase lag (up to 220°)
is observed when increasing the stimulus frequency from 0.3 Hz to
3 Hz, suggesting a more complex processing of otolith information
at the level of vestibular nuclei.

2.2.2. Intravestibular interactions
The need of including intravestibular interactions comes from
the fact that most experimental data (including the threshold and
suprathreshold studies of Meiry, 1965) are indeed obtained in
experimental conditions where intravestibular interactions occur.
In addition it is most probable, in our view, that the central processing of otolithic signals has been organized in the course of
phylogeny, as an interactive signal source in combination with
semicircular canal signals and visual signals about self motion.

Epstein (1977), Ormsby (1974) and Mayne (1974) built, although
with different methods, two models in order to account for the
perception of off-vertical rotation (rotation about an axis vertical
with respect to the head but inclined with respect to gravity).
However, Epstein's descriptive model does not succeed in explaining
perfectly the perceived constant rotation during a "barbecue" experiment. On the other hand, the model developed by Ormsby (1974)
is rather complex and cannot be presented here in detail. Roughly
speaking, this author proposes that a central processor estimates
both the "down vector" and the head angular velocity from both
otolith and canal signals. The "down vector" estimation is provided
by adding the low frequency component of the otolith signal to the
result of the integrated "down rotation" rate. The "down rotation"
rate itself is obtained by adding the high frequency component of
the semicircular canal signals to the low frequency component of
the non-conflicting canal and otoliths signals (see Fig. 6).

Although linear motion perception is not mentioned by Ormsby,
it may be inferred from the difference between the otolith signal
and the "down" estimation, assuming that the otolith system measure
is the sum of gravity ("down" estimation) and the actual linear
acceleration.

The intravestibular interactions have also been taken into
account by Mayne (1974). He proposed two different processes: one
related to the determination of the vertical, which can be compared
to the "down vector" estimation of Ormsby (1974), the other related
to the computation of the linear motion. In this model, the determination of the gravity component along a given axis is obtained by
adding the low-frequency component of the type 1 (static) otolith
cell response to the high-frequency part of the estimated rate of
change of the gravity component. This last signal is computed by
multiplying the head angular velocity (from semi-circular canals)
by the gravity component along the perpendicular axis. Once the
gravity components along the three head axis have been computed, the
direction of the vertical can be determined, and the type 2 (dynamic)
otolith cell response can be freed from the effect of gravity. The
resulting signal is then processed by three successive damped inte-
grators in order to obtain head linear acceleration, velocity and
displacement. Although the numerical values of the model constants
are given by Mayne (1974), no quantitative simulation of experimental
data has been performed.

For a complete description of these models, the reader is referred to the original publications (Ormsby, 1974; Mayne, 1974). In
spite of a rough similarity between these models, one can point out
some qualitative differences. For instance, the functional differentiation between the two main types of cell responses is not mentioned by Ormsby (1974). On the other hand, the contribution of the
otolith signal in the determination of the head angular velocity
(in a non-horizontal plane) requires stimulation of the semicircular
canals in the model of Mayne (1974). Therfore, this model cannot
explain the continuous rotation perception (and eye movements)
which is observed during constant prolonged "barbecue" rotations
about an earth horizontal axis (Guedry, 1974).

To our knowledge, no attempts have been made to correlate the
modelling of intravestibular interactions with neurophysiological
data. Canal and otolith signals are known to converge in the different vestibular nuclei, with a variable distribution of saccular,
utricular and canalicular afferent projections (see the review by
Wilson and Melvill Jones, 1979). Schor (1974) compared the responses
of descending nuclear cells to tilts in intact cats and in canal-
plugged cats: in this latter case sensitivity was strongly reduced,
suggesting a complementary role of the canals and of the otoliths
in roll or pitch movements. Additional evidence for such complementarity is provided by Anderson et al (1978) who have shown that
during roll the otoliths improve the vestibulo-ocular reflex gain
for low frequencies (below .1 Hz). The otoliths are therefore complementary to the semicircular canals. However, no such experiments
have been performed with pure translation.

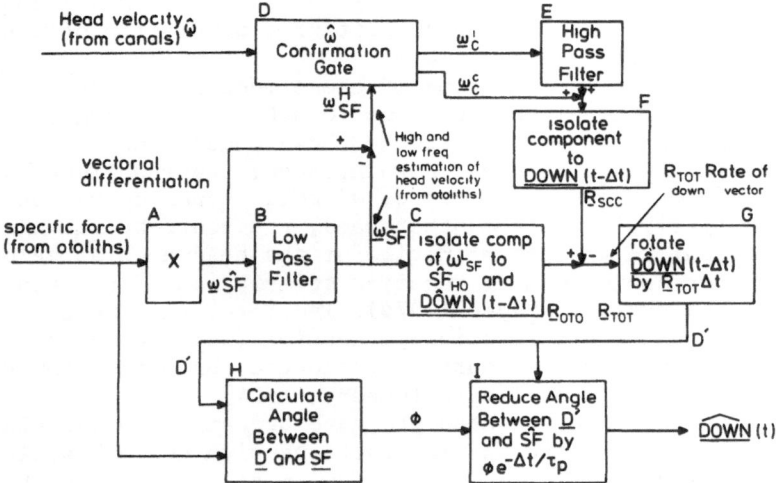

Fig. 6. Model of otolith canal interaction, (from Ormsby, 1974). The "Down" estimator.

Functional Blocks:

A: Vectorial Differentiation: computation of the angular velocity ($\omega_{\hat{S}F}$) of the specific force vector (SF) from otoliths.

B: Low pass filter: computation of the low frequency component of $\omega_{\hat{S}D}$ ($\omega_{\underline{S}F}^{L}$).

$\omega_{\underline{S}F}^{H} = \omega_{\hat{S}F} - \omega_{\underline{S}F}^{L}$ is the high frequency component of $\omega_{\hat{S}F}$.

C: Computation of R_{OTO} (rate of down vector, as estimated from otoliths informations), component of $\omega_{\underline{S}F}^{L}$ perpendicular to the plane of SF and DOWN (t−Δt).

D: Conformation Gate: Comparison of head angular velocity $\hat{\omega}$ as estimated from semicircular signals with $\omega_{\underline{S}F}^{H}$ (provided by otoliths). $\omega_{\underline{C}}^{C}$ and $\omega_{\underline{C}}^{i}$ are respectively the consistent and inconsistent portions of $\hat{\omega}$.

E: High pass filter.

F: Computation of R_{SCC} (rate of down vector, as estimated from semicircular informations). R_{TOT} is the overall rate of down vector.

G: Integration of R_{TOT} with DOWN (t−Δt) as initial value. the output of G: $\underline{D}'$ is a first estimation of the new DOWN vector.

H and I: Slow reduction of discrepancy between $\underline{D}'$ and $\underline{SF}$.

H: computation of the angle ϕ between $\underline{d}'$ and $\underline{SF}$.

I: Reduce the angle ϕ by applying the factor $e^{-\Delta t/\tau_p}$ where τ_p is a time constant.

3. THE PERCEPTION OF SELF MOTION INDUCED BY THE VISUAL SYSTEM

3.1. Physical input to the retina during linear motion: "Optic flow"

We shall first consider the properties of the physical optic stimuli impinging on the retina during linear motion. This has been called the "optic flow analysis". The optic flow on the retina does not depend upon the linear motion of the head only but is also a function of eye movements. We shall therefore successively consider the case of a subject with a paralysed eye moving in a stationary rigid world and secondly consider the case where the eyes of the subject are able to move.

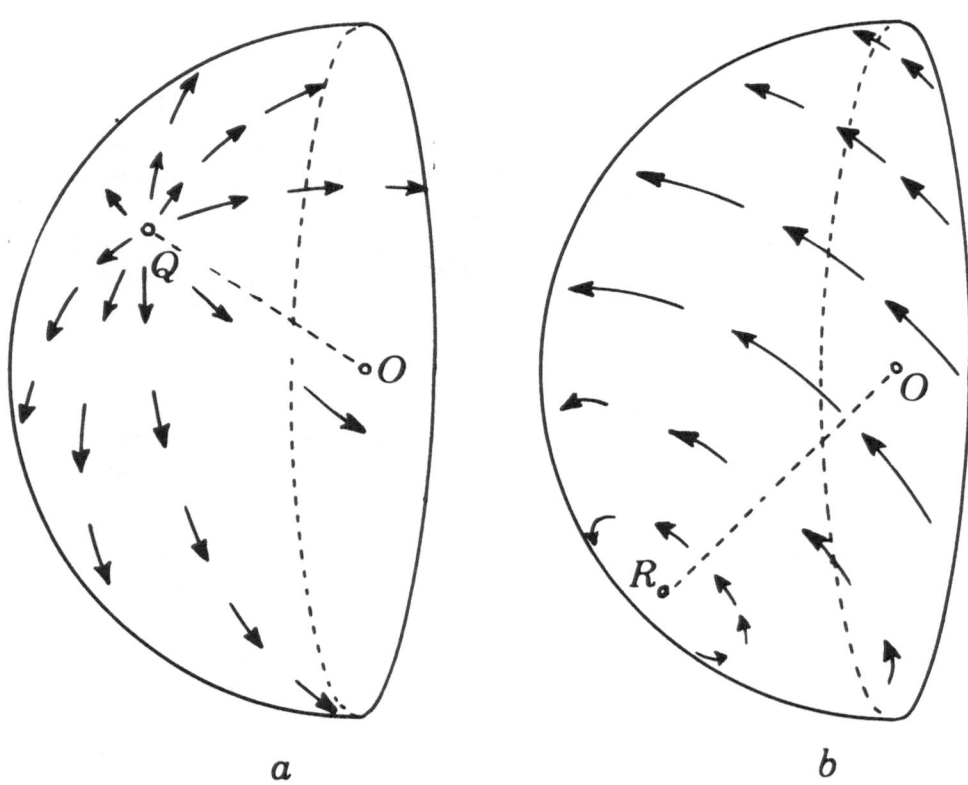

a b

Fig. 7. Typical translational ("polar", see 7a) and rotational ("axial", see 7b) components of the optical flow field on a hemispherical retina. O is the equivalent optic center of the eye. Q is the intersection of the linear motion axis with retina. R is the intersection of the angular motion axis with retina (from Longuet-Higgins and Pradzny, 1980).

3.1.1. Optic flow in a subject moving in a stationary rigid world

Let us assume that an observer is paralysed i.e. he does not make any eye movements. In the case of a pure head rotation, the retinal projection of a visual scene turns around the retinal projection of the axis of rotation. If one assumes that the retina and the visual analysing system have isotropic properties, the optic flow can be described in polar coordinates (ρ, θ) whose center is the retinal projection of the axis of rotation. This is exemplified in Fig. 7. In this case the subject's angular velocity $(\dot{H}_R)$ can be deduced from the angular velocity $\dot{\theta}$ of any texture element projecting on the retina:

$$\dot{H}_R = \dot{\theta}$$

In the case of pure translation, as analysed by Gibson (1954, 1957) and Hay (1966) the optic flow can be defined as a field of vectors which are magnified (or decreased) from a pole which is the retinal projection of the axis of translation. In similar polar coordinates (ρ, θ) the head linear velocity $(\dot{H}_T)$ can be deduced from the radial velocity and depth (D) of any surround element projecting on the retina:

$$\frac{\dot{H}_T}{D} \cong \frac{\dot{\rho}}{\rho}$$

where $\dot{\rho}$ is the rate of change of the radius ρ. Note that this formula has the general form of Weber's law. Note also the inverse simple relationship:

$$\frac{D}{\dot{H}_T} \cong \frac{\rho}{\dot{\rho}} = \frac{1}{\text{(Rate of dilation of retinal image)}}$$

as used by Lee (1974) in his theory concerning the visual control of braking. $\frac{\rho}{\dot{\rho}}$ can be extracted easily from any retinal image and may be interpreted as a "time to collision" estimation. As pointed out by Longuet-Higgins and Pradzny (1980), the distinction between rotational and translational components of movement using only visual motion information is not an easy task, even for a perfect algebrist located somewhere in the nervous sytems. (The same difficulties have been noted above concerning otolithic information.) These authors have postulated that such a distinction can be made by the utilisation of motion parallax (Helmholtz, 1925). Similar concepts were used by Gibson (1966) in his use of the terms "obstacle" and "opening". An "obstacle" is specified by a loss of the moving visual structure inside a closed contour. Loss or gain of seen structure implies discontinuities in the local retinal image velocity due to different depths or localisations of objects projecting on the same retinal locality. Since these discontinuities only occur during translations and not during rotations, they could be involved in the visual discrimination between rotation and translation.

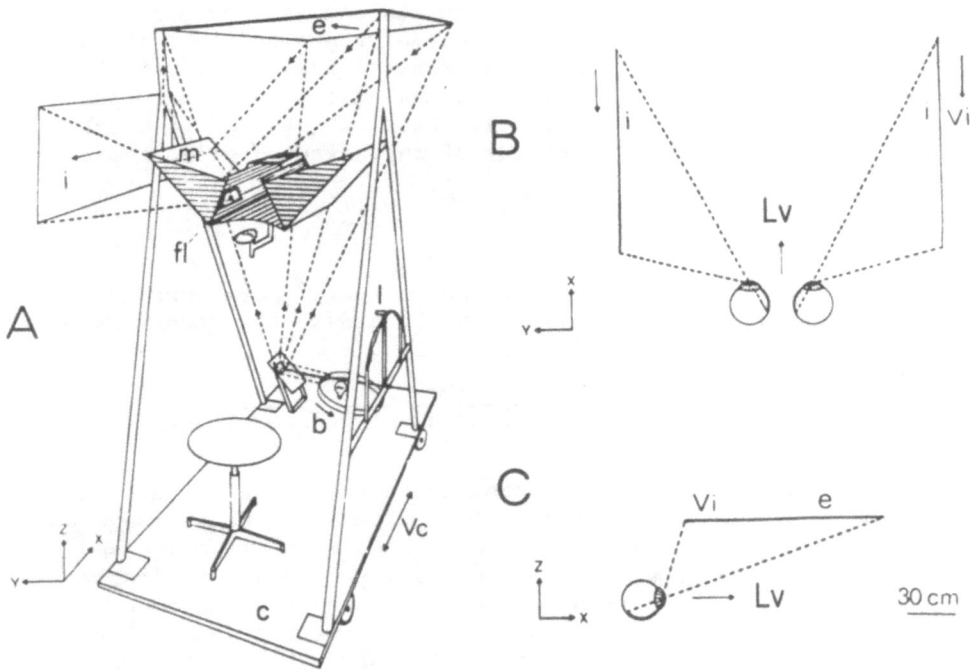

More generally we may suggest (Droulez, in preparation) that
the translational and rotational components of movement may be separated by a local partial derivative of retinal image velocity expressed in retinal coordinates (for instance ρ and θ). The case of
discontinuity is then a particular case in which the partial derivative of retinal image velocity takes an extreme value. As pointed
out by Longuet-Higgins and Pradzny (1980), visual linear velocity
estimation has to be scaled in terms of depth estimation. In other
words, the retinal image velocity can only provide information about
the linear velocity/depth ratio. Therefore stereopsis, retinal
images of objects of a known size, relative magnitudes of visual
objects and motion parallax, together with vestibular and proprioceptive information may all be involved in this depth estimation
during linear motion. It is somewhat surprising that even in the
case of pure rotation, depth estimation modifies circular velocity
estimation (see Wist et al., 1975; and Dichgans and Brandt, 1978 on
the Pülfrich effect). Such a phenomenon may be the consequence of
imperfect linear-circular visual discrimination.

3.1.2. Role of eye movements

Although a number of experimental results have been obtained
concerning the influence of eye movements on circular vection, there
is at present practically no information concerning the role of eye

Fig. 8. Experimental apparatus.
A - The subject is seated on a cart (c). A band of film (b),
on which are printed randomly distributed signs (letters,
points, crosses, etc.) is moved by a projector. The image of
this band is projected on a screen (e) (surface 1,6 m², distance to the horizontal plane of the eyes 0,5 m). The subject, whose head is maintained in a normal upright position
by a chin rest, looks through three openings in a box: two
lateral windows (fl), and an opening above the head. Two
virtual images (i) of the screen (e) are given by two mirrors
(m) tilted at 45° angles. The subject is required to give
continuous magnitude estimations of his subjective sensation
of linear-vection (LV) when the film is in motion at the
speed V_i, or when the cart is moving at the speed V_c, by
using the hand lever (l) on his right side. This lever is
also used to control V_i, in the active procedure.
B - Schematic view of image position (i) from above, with
respect to the subject's eye (angle of vision: 20° to 70°).
LV indicates direction of subjective velocity (forward LV
for backward V_i).
C - View from the right indicating the position of the screen
(e). Same notation as in A and B.
(From Berthoz et al., 1975.)

movements in linear motion perception (see however, Buizza et al., 1979, 1980).

It is therefore quite urgent for experiments to be done in this field. The fact that eye movements add a rotary component to the perceived translation introduces some complication. In addition, in the case of a translation, the *distance* at which a visual stimulus is presented will be an important factor affecting eye movements. Therefore, depth and linear motion are interactive factors in such cases.

3.2. Linear Vection

3.2.1. Methods

By linear vection (LV) we define a perception of linear self motion induced by a moving visual scene. The occurence of this self motion perception, which is different from the perception of outside world motion, is exemplified by the "train illusion": when a train next to the train in which an observer is seated begins to move, the observer often has the illusion that it is the train in which he is seated that is moving.

Various types of experimental devices have been designed to study linear vection estimation. Salvatore (1968) has studied the contribution of peripheral and central vision in a laboratory experimental car using a mask occluding parts of the visual field. Denton (1966) has used in the laboratory the projection of films with an optical device and a servocontrolled camera which allowed the experimenter to simulate velocities up to 300 km/h with a field depth of 67 m. Our own device (Berthoz et al., 1974, 1975) was built in order to combine and manipulate independently visual and vestibular linear motion (Fig. 8).

3.2.2. Basic features of linear vection (LV)

The basic features of LV have been studied by Berthoz et al. (1974, 1975) for horizontal and by Chu (1976) in Young's laboratory for vertical motion using peripheral linear visual motion stimulation. These authors described the following properties:

The *peripheral visual field* is certainly essential for the appearance of LV. When velocity thresholds were measured horizontal LV was shown to appear at *image velocities* around 0.03 m/s with an image at about 0.5 m lateral from the subject's eye. Thresholds for backward and downward LV were lower than for forward and upward LV. LV increases with image velocity, but when image velocity increases further, *saturation* and then a *loss of vection* can be observed. Note that a velocity threshold and saturation are also observed in circular vection (Leibowitz and Dichgans, 1977).

On first exposure to a moving visual stimulus often *a long delay* occurs before the onset of LV (sometimes up to 20 s.). Subsequent average latencies were found to be about 1 to 2 sec for horizontal LV (Berthoz et al., 1974) with image linear velocities between 0.2 and 1 m/sec, and 2 to 3 sec for vertical LV (Chu, 1976). These

values are comparable to those obtained by Brandt et al (1973) for
circular vection (from 1 to 10 sec depending on the subject's sensitivity).

A frequency analysis of horizontal and vertical LV was performed by Berthoz et al. (1974) and Chu (1976). Figure 9 shows a summary
of the data which seem to suggest a cut off point around .5 Hz. A
comparison between these data and the dynamics of the influence of
visual scene motion on postural control has been given in Berthoz
et al. (1979).

The *luminance thresholds* for the appearance of vection have
also been measured. They are very close to the absolute luminance
thresholds for image detection by the visual system (about 2.10^{-4}
Cd/m^2 with image velocities between 0.2 and 1 m/sec). Denton (1966,
1971), studying velocity estimation during simulated car driving
measured the influence of *spatial frequency* (or image density) expressed in cycles per degree (or number of visual elements per unit
of surface). He showed that linear vection increased with spatial
frequency. Berthoz obtained unpublished data suggesting a similar
influence of spatial frequency on LV. Confirmation of this result
was obtained by Lestienne et al. (1977) who studied the effect of
moving visual scenes on posture: postural readjustment was found to
be related to the logarithm of spatial frequency. Diener et al.
(1976) showed a linear relationship between image motion perception
and spatial frequency, however, this influence has not been found
in circular vection (Dichgans and Brandt, 1978).

It is known from vehicle driving experiments that exposure to
a long duration constant velocity leads to underestimation of vehicle
velocity (Denton, 1971; Salvatore, 1968; Schmidt and Tiffin, 1969).
Irving (1973) studied adaptation to visual linear velocity in real
conditions: the driver of a car could control the velocity of his
vehicle in order to maintain his vection constant. In these conditions an adaptation was observed with a time constant of about
50 sec. Similar time constants were obtained by Berthoz et al. (1974)
in their experiments. These values are indicative of a short term
adaptation. However, there may also exist a longer time constant
adaptation which has not been studied although it should be: adaptation is considered to be responsible for many car accidents
(Desrosiers, 1967). The factors influencing adaptation are numerous,
however. For the best estimated velocities (50 to 60 km/h) the adaptation is small even if stimulation duration is as long as 20 min
(Barch, 1958; Schmidt and Tiffin, 1969). Denton (1966) has proposed
a psychometric curve describing the time course of adaptation for
various initial velocities.

3.3. Modeling of linear vection

The title of this section should have been the lack of modeling
of linear vection. In comparison to the vestibular organs, the input
to the visual system is not well defined: An analysis of the influence of the various stimulation parameters rather reflects the

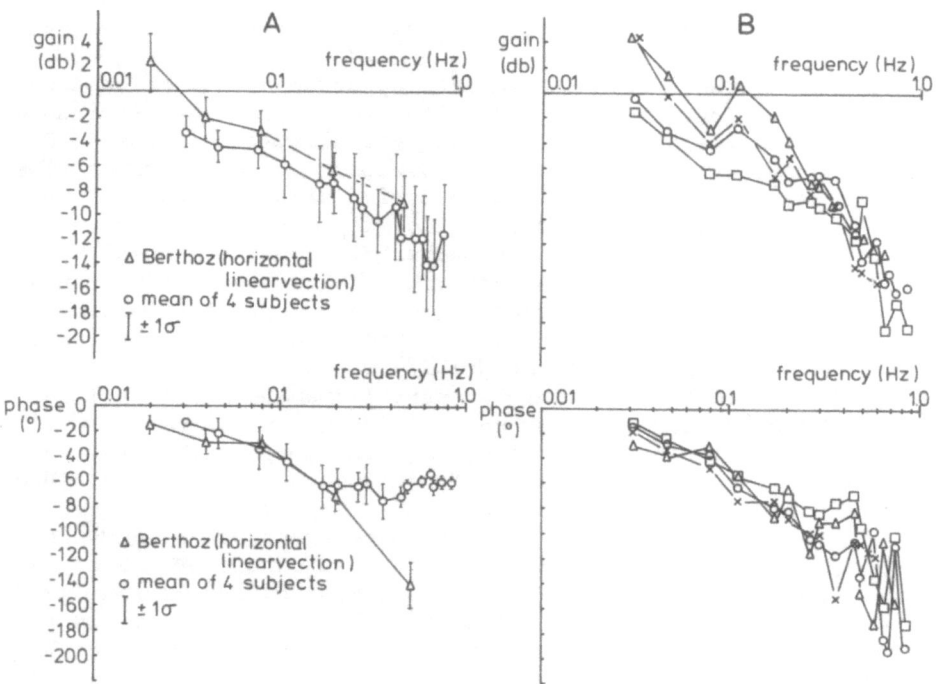

Fig. 9. Gain and phase plots of linear vection (LV).
A – Vertical and horizontal vection: Gain and phase of the
magnitude estimation of LV induced by stripes moving with
different frequencies. The data (mean values of four subjects) obtained by Chu (1976) for vertical LV (circles) are
compared with those of Berthoz et al. (1975) for horizontal
LV (triangles).
B – Vertical linear vection: frequency response to moving
stripes with a pseudo-random velocity profile (four subjects, means for each subject, each one with a different
symbol). Data from Chu (1976).

flexibility of the experimental device than that of the visual system. Moreover, the organ of vision subserves numerous functions, especially in mammals. Among them, image motion perception and vection are only partial aspects and should be integrated into a global model of vision. However, one can avoid these difficulties by reducing the input parameters to those directly related to self motion: axis of motion, velocity and time course. In this way, we neglect the influence of other parameters like depth, the horizontal and vertical extent of the image, its spatial frequency, luminance, colour, and so on, which are then considered as contingent to a particular experimental set up. Furthermore, we leave unexplored that part of the central nervous (visual) system which distinguishes between image deformation and motion of an image, between object motion and self motion, between linear and circular movements, and so on. Given such an oversimplification (which has been used for modeling oculomotor control by vision), (e.g. Raphan et al., 1977; Robinson, 1977), the interest of such a model is to describe the dynamics of one particular function of the visual system in formulations which allow the analysis of visual vestibular interactions.

The dynamics of linear vection inferred from data of Berthoz et al. (1974) and Chu (1976) can be expressed in terms of a low pass filter with a time constant of about 1 sec (Pavard and Berthoz, 1977). However, one should keep in mind the asymmetry (backward and downward LV are more sensitive than forward and upward LV) and the non-linear characteristics of LV (threshold, saturation).

4. VISUAL VESTIBULAR INTERACTION IN LINEAR SELF MOTION PERCEPTION

4.1. Influence of otolithic stimulation on eye movements

4.1.1. Eye movements in the dark

Linear accelerations applied to the head have been proven to induce various types of eye movements in the dark. These compensatory eye movements, together with those induced by semicircular canals stimulations, help to stabilize the visual world during natural head movements. During rotations about an off-vertical axis, the canal and otolith ocular reflexes cooperate in order to induce torsional eye movements about the axis or vertical eye movements in pitch (about the Y axis) or horizontal nystagmus in barbecue rotations (about a horizontal Z axis). Pure linear accelerations along the Y axis and the Z axis can evoke horizontal nystagmus: the so-called L-Nystagmus (Jongkees and Philipzoon, 1962; Niven et al., 1966; Young, 1967) and vertical nystagmus (McCabe, 1964).

The influence of otoliths on eye movements in the dark can be studied by indirect methods, for example comparing the gain of the vestibulo-ocular reflex during rotations about the vertical and off-vertical axis (see for a review: Benson, 1974) and by direct methods using pure otolithic stimulation (Jongkees and Philipzoon, 1962; Niven et al., 1966; McCabe, 1964; Buizza et al., 1979, 1980).

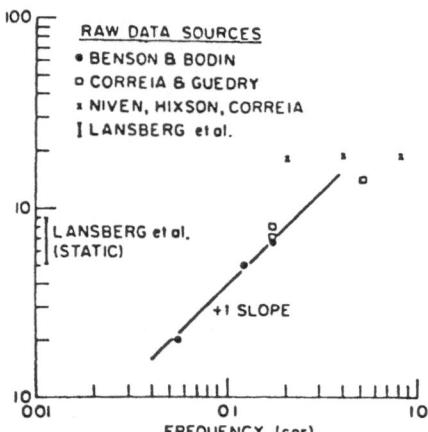

Fig. 10. L-Nystagmus sensitivity, expressed as the horizontal slow
         phase eye velocity (deg/sec) normalized to a 1 g stimulus,
         versus frequency of linear horizontal oscillation (Hz).
         Data collected by Young (1967) from different authors.

Constant rotations about a vertical Z axis evoke a horizontal
nystagmus, the slow phase velocity of which decays regularly to 0
with lower frequencies (time constant about 15 sec). Constant rotation about the same Z axis but oriented perpendicular to gravity
("barbecue" condition) evokes in addition a permanent "barbecue"
nystagmus with a sinusoidal component and a bias component (Benson
and Bodin, 1966; Bodin, 1968; Benson, 1974). These two components
may originate from otolithic stimulation because specific destruc-
tion of the utricular nerve entails the suppression of sustained
nystagmus (Janeke, 1968) while the blockage of all six semicircular
canals does not alter it (Correia and Money, 1968). However, the
mechanism of these responses is still unclear. Goldberg (1981)
showed that in the squirrel monkey persistent nystagmus is not based
on a mechanical response from the semi-circular canals.

     When subjects are submitted to horizontal linear oscillations
along the Y axis, they exhibit horizontal nystagmic eye movements.
Their slow phase velocity is closely correlated to the actual linear
acceleration of the head with a sensitivity of about 15°/sec/g at
0.2 Hz (Niven et al., 1966; Buizza et al., 1980), (see Fig. 10).
This value can be compared with the peak to peak sinusoidal compo-

nent in the nystagmus caused by "barbecue" rotations (Benson and Bodin, 1966; Correia and Guedry, 1966) of about 20°/sec/g at 0.2 Hz. Levison and Zacharias (1978) suggested that the bias component of "barbecue" nystagmus may result from a central treatment of oto-lithic information in a manner similar to the way in which the per-ception of rotation can be deduced from a rotating linear accelera-tion, like in Ormsby's model (1974). In other words, the otolith organ seems to be involved in perception and in compensatory eye movements during head rotations as well as during head translations. For the sake of comparison, the gain of the otolithic influence or horizontal eye movements as deduced from "barbecue" experiments (Benson and Bodin, 1966) is: 0.5 at 0.0028 Hz, 0.6 at 0.055 Hz and 0.27 at 0.16 Hz.

4.1.2. Eye movements in the light

Although the above mentioned eye movements were recorded in the dark, the otolithic influence on eye movements is assumed to play a role in the ocular compensation of head movements in the light together with other compensatory eye mechanisms, particularly the optokinetic nystagmus. For the case of the semicircular canals visual-vestibular interaction was studied by Koenig et al. (1978) who showed that vestibular and visual influences on eye movements during rotations about a vertical axis are algebrically added. The optokinetic nystagmus is dominant in the low frequency range while the vestibulo-ocular reflex is more essential in the high frequency range. In the field of otolith visual interaction in man, Tokunaga (1977) and Buizza et al. (1980) demonstrated an important modifica-tion of optokinetic nystagmus when subjects are submitted to horizon-tal linear oscillations (Fig. 11). The main feature of these modifi-cations is a sinusoidal modulation of the optokinetic slow phase velocity which may be closely related to otolith stimulation. How-ever, the peak to peak amplitude of this modulation is much higher (4 to 5 times) than what can be calculated assuming a simple addi-tion between optokinetic and "L" Nystagmus. Moreover this modulation appears at a very low level of linear acceleration: 0.1 m/sec², which is close to the perception threshold (see Table I). These findings suggest a complex treatment of otolith information in the control of eye movements, involving multisensory interactions.

4.2. Influence of linear acceleration on perceived image velocity

It is known from car driving studies that linear acceleration may produce modifications of perceived self velocity in drivers. In his study, Salvatore (1968) showed that, as mentioned above, drivers could scale rather well their velocity in terms of car velocity when there was no acceleration. He showed that transient accelerations degraded this estimation. In the laboratory, using their moving cart, Berthoz et al. (1974, 1975) showed qualitatively and quantita-tively that when vestibular and somatosensory cues are in conflict with visual image motion in the fore-aft directions (X axis), LV

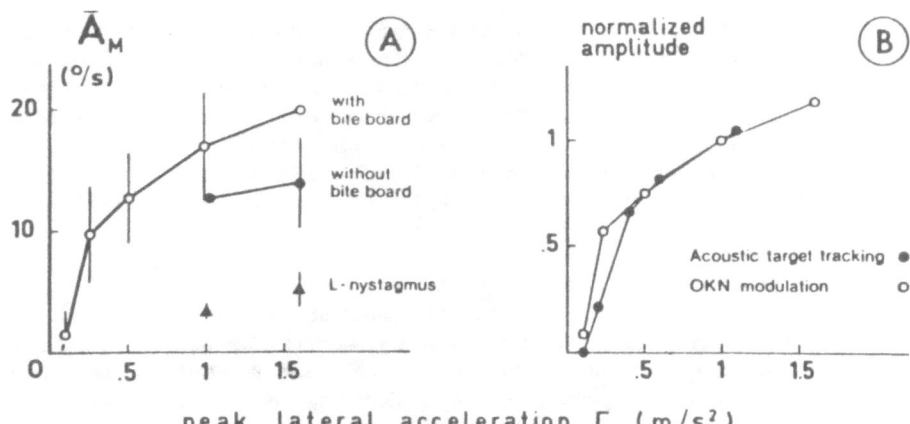

Fig. 11. Influence of linear acceleration on optokinetic nystagmus. Relationship between slow phase velocity and lateral acceleration.

A – The mean amplitude $\overline{A}_M$ of slow phase velocity modulation increased non-linearly with peak acceleration, with a threshold corresponding to the known threshold for motion detection by the otoliths. The results obtained when a bite-board was used to improve head fixation (open circles) are compared with those obtained without bite-board (filled dots) and with the peak to peak amplitude of the L-nystagmus (triangles). Mean values and one standard deviation from 7 subjects (11 for L-nystagmus). Data without standard deviation refer to control trials performed on two subjects only. Parameters of the optokinetic stimulation: amplitude of the horizontal visual angle: 90°, velocity: 12°/sec.

B – In a preceding work (Buizza et al., 1979) the relationship between the amplitude of the slow components of the eye movements during tracking of acoustic targets and the level of subject lateral acceleration was found. This curve (filled dots) is plotted here together with the data shown in A (open circles) for the sake of comparison. Each set of data has been normalized with respect to the value corresponding to 1 m/sec².

can be drastically reduced and striking errors can be introduced in
the subjective evaluation of motion direction. On the one hand they
demonstrated that when constant LV is induced by a visual scene
moving with a constant velocity at the periphery along the X axis,
subjects cannot correctly detect cart motion (there is a dominance
of vision over vestibular cues). On the other hand, during transient
acceleration, the subject loses vection on some occasions (see
Fig. 12). These results demonstrate that visual-vestibular interactions indeed occur during linear motion.

The phenomena were studied in detail by Pavard and Berthoz
(1977) using the same experimental device (see Fig. 8). In a first
experiment, they gave pseudo-random movements to the cart along the
X axis. Image velocity was set at an initial value, but subjects
were asked to control image velocity so as to keep it constant. Two
changes were observed: a) subjects slowly increased image velocity
with time, this being due to adaptation and, b) during forward or
backward acceleration (triangles of velocity) the subjects perceived
a transient decrease of image velocity and therefore increased the
actual image velocity. A subsequent quantitative investigation in a
second experiment revealed the following fact: the perceived slowing
down or even the subjective stabilization of image velocity (which
in this experiment was kept constant with respect to the subject)
was found to be proportional to the amplitude of cart acceleration.
For accelerations between 0 en 0.5 m/sec² total subjective stabilization of the visual scene was perceived. The duration of this stabilization lasted from 0 to 2.5 seconds. The duration decreased when
image velocity increased for a given acceleration (see Fig. 13). It
was greatest when image and cart velocity were in opposite directions
(Natural conditions). When image and cart velocities were in the same
direction (conflict) subjects could not analyse precisely their perceptions. This visual stabilization was also found to be dependent
upon the *degree of colinearity* between image and cart velocity, and
upon spatial frequency (increased spatial frequency was accompanied
by a subjective increase of stabilization time).

The results demonstrate that, in the absence of eye movements,
linear acceleration can modify drastically the perceived velocity of
a visual scene. However, this effect is dependent upon the relative
weighing between visual and vestibular input. For small image velocities (0.4 m/sec) accelerations induced an underestimation of image
velocity which can lead to a perceived complete stabilization of the
visual world. For image velocities between 0.5 and 0.8 m/sec depending upon whether image and cart velocities where of opposite direction or in the same direction, either an underestimation or an over-
estimation of image velocities could be seen. For high image velocities the acceleration of the cart could not further modify image
motion perception. The visual input is strongly dominating (saturation of LV) and the vestibular signals cannot appreciably modify
visual information. This stabilization was related by Pavard and
Berthoz (1977) to the dynamics of the otoliths. A simple model was
proposed in order to predict these results (see Fig. 14).

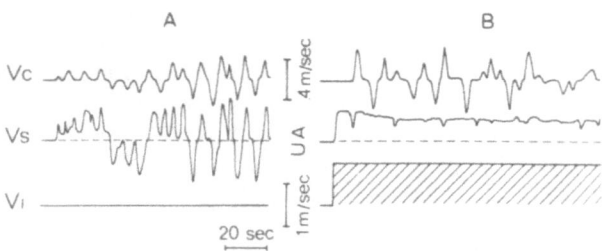

Fig. 12. Interaction of subjective velocity sensation (V_s) induced
by conflicting motion of the visual scene and by the translational velocity (V_c) of the cart.

A – Vestibular detection of linear motion with eyes closed.
From top to bottom: cart velocity (V_c) subjective velocity
(V_s)-(arbitrary units) and image velocity (V_i). Note that
the subject detects the variations of cart velocity well,
with a slight delay.

B – Conflict between a constant velocity visual scene and
oscillating cart velocity. From top to bottom: cart velocity, subjective velocity (combination of vestibular and
visually induced motion sensation (UA) and image velocity
(which is constant, equal to 0.5 m/sec moving backward and
inducing a forward linear vection. Note that the subject
indicates a subjective velocity which is always forward,
even during periods when the cart is moving backward (from
Berthoz et al., 1974).

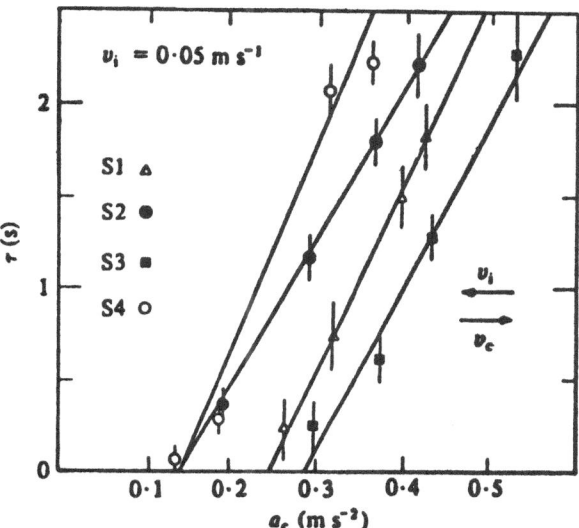

Fig. 13. Influence of linear acceleration on the perceived velocity
         of a visual pattern moving at constant velocity with re-
         spect to the observer.

         Duration of stabilization time τ(s) plotted as a function
         of cart acceleration (a_c). Results are shown for four
         subjects (S1, S2, S3 and S4). Image velocity (v_i) was main-
         tained constant relative to the observers. The bars re-
         present standard deviation (N = 10). (From Pavard and
         Berthoz, 1977.) See text.

4.3. Modeling visual vestibular interactions

As mentioned above, the lack of relevant models of the visual
system renders the problem of modeling visuo-vestibular interactions
rather hazardous in the case of linear motion. With the assumption
that the visual system can perform a visual estimation of eye motion
in the world (or conversely world motion relative to the eye), some
models have been developed either in the field of eye movement
control (see Robinson, 1977; Raphan et al., 1977) or in the field

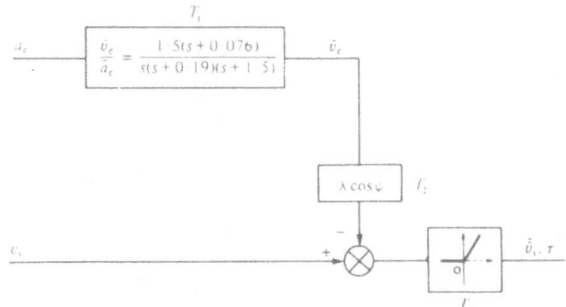

Fig. 14. Model. Simplest model of vestibulo-visual interaction
        allowing the test of some mechanisms for the perception of
        movement. T_1 is the transfer function of the vestibular
        system: S is complex frequency (Young and Meiry, 1968).
        The vestibular sensation of the movement (v_c) induced by
        the cart acceleration (a_c) is combined geometrically with
        image velocity (v_i) after block T_2. ψ is the angle between
        the direction of the two motions (visual and vestibular).
        A nonlinear transformation T_3 is necessary to take into
        account experimental results. The visual perception of
        movement $\tilde{\tilde{v}}_i$ and the time stabilization (τ) are computed
        from this model. λ is an adjustable parameter of the model.
        (From Pavard and Berthoz, 1977.)

of self motion perception (see review by Henn et al., 1980) during
head rotations. However, little has been done in the field of otolith visual interactions. Buizza et al. (1980) developed a model of
otolith-optokinetic interactions. The model predicts the slow phase
velocity of nystagmus in the dark with various combinations of otolithic and optokinetic stimulations.

In this model, the otolithic estimation of head linear velocity
is added to the output of the optokinetic system (Schmid et al.,
1979) via a gain control operator. This gain control was included
in order to predict the implementation of otolithic influence on eye
movements when subjects receive visual information. It has been suggested that a visual estimation of depth (distance between visual

target and subject's eyes) might play a role in the otolithic gain control, due to the fact that an adequate eye compensation of a head translation must depend upon the target distance.

Pavard and Berthoz (1977) developed a somewhat similar additive model in order to simulate the modification of perceived image velocity induced by linear acceleration (see Fig. 14). The otolithic model was chosen identical to the "revised model" proposed by Young and Meiry (1968). The output of the otolithic system is added to a visual estimation of image velocity via a gain control operator which is dependent upon the angle between the actual and the visual linear displacement. A first order non-linearity was also introduced to take into account the fact that in the experimental conditions vestibular stimulation degrades visual information.

REFERENCES

Adrian, E.D., Discharges from vestibular receptors in the cat. J. Physiol. (London), 1943, 101, 389-407.

Anderson, J.H., Blanks, R.H.L. and Precht, W., Response characteristics of semicircular canal and otolith systems in cat. I: Dynamic responses of primary vestibular fibers. Exp. Brain Res. 1978, 32, 491-507.

Baker, R. and Berthoz, A., Control of gaze by brainstem neurons. Developments in Neuroscience (Vol. 1), Amsterdam: Elsevier, 1977.

Barch, A.M., Judgements of speed on the open highway. Journal of Appl. Psychol., 1958, 42, 362-366.

Barlow, J.S., Inertial navigation as a basis for animal navigation. Journal Theor. Biol., 1964, 6, 76-117.

Békésy, G. von, Uber die Stärke der Vibrationsempfindung und ihre objektive Messung. Sonderdruck aus "Akustische Zeits", 1940, 5, 113-124.

Benson, A.J., Modification of the response to angular accelerations by linear accelerations. In: Handbook of Sensory Physiology, Vol. VI/2: Vestibular System, H.H. Kornhuber (Ed.), Springer-Verlag, New York, 1974, 281-320.

Benson, A.J. and Bodin, M. Interaction of linear and angular accelerations on vestibular receptors in man. Aerospace Med., 1966, 5, 113-124.

Benson, A.J., Diaz, E. and Farrugia, P., The perception of body orientation relative to a rotating linear acceleration vector. Fortschr. Zool., 1975, 23, 264-274.

Berthoz, A., Rôle de la proprioception dans le contrôle de la posture et du geste. In: H. Haécan and M. Jeannerod (Eds.), Du contrôle monteur à l'organisation du geste, Paris, Masson, 1978.

Berthoz, A., Pavard, B. and Young, L.R., Rôle de la vision périphérique et interactions visuo-vestibulaires dans lan perception exocentrique du mouvement linéaire chez l'homme. C.R. Acad.

Sci. (Paris), 1974, <u>278</u>, D, 1605-1608.

Berthoz, A., Pavard, B. and Young, L.R., Perception of linear hori-
zontal self motion induced by peripheral vision (linear vection).
Basic characteristics and visual vestibular interaction. <u>Exp.
Brain Res.</u>, 1975, <u>23</u>, 471-489.

Berthoz, A., Lacour, M., Soechting, J.F. and Vidal, P.P. The role
of vision in the control of posture during linear motion.
In: R. Granit and O. Pompeiano (Eds.). <u>Control of posture and
movement</u>. Progress in Brain Research: (Vol. 50), Amsterdam,
Elsevier, 1979.

Bodin, M.A., The effect of gravity on human vestibular responses
during rotation in pitch. <u>Journal Physiol</u>. (London), 1968,
<u>196</u>, 74-75.

Brandt, Th., Dichgans, J. and Koenig, E., Differential effects of
central versus peripheral vision on egocentric and exocentric
motion perception. <u>Exp. Brain Res.</u>, 1973, <u>16</u>, 476-491.

Brandt, Th., Büchele, W. and Arnold, F., Arthrokinetic nystagmus
and egomotion sensation. <u>Exp. Brain Res.</u>, 1977, <u>30</u>, 331-338.

Buizza, A., Léger, A., Berthoz, A. and Schmid, R., Otolithic acoustic
interaction in the control of eye movement. <u>Exp. Brain Res</u>,
1979, <u>36</u>, 509-522.

Buizza, A., Léger, A., Droulez, J., Berthoz, A. and Schmid, R.,
Influence of Otolithic Stimulation by Horizontal Linear Acceleration on Optokinetic Nystagmus and Visual Motion Perception.
<u>Exp. Brain Res</u>, 1980, <u>39</u>, 165-176.

Chu, W.H.N., Dynamic response of human linear vection. S.M. Thesis,
MIT, Cambridge, Mass., 1976.

Correia, M.J. and Guedry, F.E., Modification of vestibular responses
as a function of rate of rotation about an Earth-horizontal
axis. <u>Acta oto-laryng</u>, (Stockholm), 1966, <u>62</u>, 297-308.

Correia, M.J. and Money, K.E., The effect of blockage of all six
semicircular canal ducts on nystagmus produced by dynamic
linear acceleration in the cat. DRET Report 728. Toronto:
Defence Research Establishment, Defence Research Board,
1968.

Daunton, N.G. and Thomsen, D.D., Otolith-visual interaction in
single units of cat vestibular nuclei. <u>Neuroscien. Abstr.</u>, II
1976, <u>2</u>, 1526.

Denton, G.G., A subjective scale of speed when driving a motor
vehicle. <u>Ergonomics</u>, 1966, <u>9</u>, (3), 203-210.

Denton, G.G., The influence of visual pattern on perceived speed.
Road Res. Lab., Crowthorne Report L.R., 1971, <u>409</u>, 1-7.

Desrosiers, R.D., Speed estimation on residential streets. <u>Public
Roads</u>, 1962, <u>32</u>, 74-76.

Dichgans, J. and Brandt, Th., Visual vestibular interaction: Effects
on self-motion perception and in postural control. In: Handbook of Sensory Physiology, Vol. VIII, R. Held, H.W. Leibowitz,
and H.L. Teuber (Eds.). <u>Perception</u>, Springer, Berlin, Heidelberg, New York, 1978, 755-804.

Diener, H.C., Wist, E.R., Dichgans, J. and Brandth, Th., The spatial-frequency effect on perceived velocity. Vision Res., 1976, 16, 169-176.

Dodge, R., Thresholds of rotation. J. Exp. Psychol, 1923, 6, 107-137.

Einstein, A., The Meaning of Relativity. Princeton University Press, Princeton, N.J., 1945.

Epstein, L.I., On the Interaction of Otolithic and Cupular Sensations. Aviation Space Environ. Med., 1977, 48, 200-202.

Fernandez, C. and Goldberg, J., Physiology of Peripheral Neurons Innervating Otolith Organs of the Squirrel Monkey (I, II and III). J. Neurophysiol., 1976, 39, 970-1008.

Fischer, M.H. and Kornmüller, A.E., Optokinetisch ausgelöste Bewegungswahrnehmung und optokinetischer Nystagmus. J. Psychol. Neurol., (Lpz), 1930, 41, 273-308.

Gibson, J.J., The visual perception of objective motion and subjective movement. Psychol. Rev., 1954, 61, (5), 304-314.

Gibson, J.J., The senses considered as perceptual systems. Houghton Miffin, Boston, 1966, 336 pp.

Gibson, J.J. and Gibson, E.J., Continuous perspective transformations and the perception of rigid motion. J. Exp. Psychol., 1957, 54, 129-138.

Goldberg, J.M. and Fernandez, C., Eye Movements and Vestibular-Nerve Responses Produced in the Squirrel Monkey by Rotations about an Earth-Horizontal Axis. Proceeding of Barany's Society, 1980. In press, 1981.

Graybiel, A., Measurement of otolith function in man. In: H.H. Kornhüber (Ed.), Handbook of Sensory Physiology. Vol. VI/2. Vestibular system. Tasks, Springer, Berlin, Heidelberg, New York, 1974, 233-266.

Guedry, F.E., Psychophysics of vestibular sensation. In: H.H. Kornhüber (Ed.), Handbook of Sensory Physiology. Vol. VI/2. Vestibular system. Task, Springer, Berlin, Heidelberg, New York, 1974.

Guedry, F.E. and Harris, C.S., Labyrinthine function related to experiments on the parallel swing. NSAM-874. Pensacola, Fla.: Naval School of Aviation Medicine, 1963.

Gundry, M.J., Threshold of perception for periodic linear motion. Aviation Space Environ. Medicine, 1978, 49, 679-686.

Gurnee, H., Thresholds of vertical movement of the body. J. Exp. Psychol., 1934, 17, 270-285.

Hay, J.C., Optical motions and space perception: an extension of Gibson's analysis. Psychol. Rev., 1966, 73, 550-565.

Helmholtz, H. von, Handbuch der physiologische Optik. Voss, Hamburg Leipzig, 1896.

Helmholtz, H. von, Treatise on physiological optics. J.P.C. Southall (Ed.). Dover, New York, 1925.

Henn, V., Cohen, B. and Young, L.R., Visual-vestibular Interaction in Motion Perception and the Generation of Nystagmus. Neurosciences Res. Prog. Bull., 1980, 18, 4.

Hixson, W.C., Niven, J.I. and Correia, M.J., Kinematics nomenclature for psychological accelerations. Monograph 14. Pensacola, Fla.: Naval Aerospace Medical Institute, 1966.

Hosman, R.J.A.W., and Vaart, J.C. van der, Vestibular models and thresholds of motion perception. Results of tests in a flight simulator. Delft, University of Technology, Dept. of Aerospace Engineering, Report LR-265, 1978.

Irving, A., The perceptual problems of the driver. Report of the 1st Congress of I.D.B.R.A., Zürich, 1973.

Janeke, J.B., On nystagmus and otoliths. (A vestibular study of responses as provoked by cephalo-candal horizontal axial rotation.) Thesis Cloeck en Moedigh, Amsterdam, 1968.

Jones, G.M. and Young, L.R., Subjective detection of vertical acceleration: A velocity dependent response DRB Aviat. Med. Res. Unit Rep. 5, 245-255; Report No. DR 225, 1976.

Jongkees, L.B.W. and Groen, J.J., The nature of the vestibular stimulus. J. Laryng., 61, 529-541, 1946.

Jongkees, L.B.W. and Groen, J.J., A quantitative analysis of the reactions of a person after loss of the function of both inner ears. J. Laryng., 64, 135-140, 1950.

Jongkees, L.B.W. and Philipszoon, A.T., Nystagmus provoked by linear accelerations. Acta Physiol. Pharmacol. Neerl., 10, 238-247, 1962.

Kellogg, R.S., Dynamic counterrolling of the eye in normal subjects and in persons with bilateral labyrinthine defect. NASA SP-77, Washington, D.C., NASA 1965.

Koenig, E., Allum, J.H.J and Dichgans, J., Visual vestibular interaction upon nystagmus slow phase velocity in man. Acta Otolaryngol., 85, 397-410, 1978.

Lansberg, M.P., Some considerations and investigations in the field of labyrinthe functioning. Aeromed. Acta, 3, 209-219, 1954.

Lee, D.N., Visual information during locomotion In: McLeod and Pick (Eds.), Perception. Cornell University Press, Ithaca, 250-267, 1974.

Leibowitz, H.W. and Dichgans, J., Zwei verschiedene Seh-Systeme. Umschau in Wissenschaft und Technik, 77, (1), 353-354, 1977.

Lestienne, F., Soechting, J. and Berthoz, A., Postural readjustments induced by linear motion of visual scenes. Exp. Brain Res., 28, 363-384, 1977.

Levison, W.H. and Zacharias, G.L., Motion cue models for pilot vehicle analysis. Aerospace Medical Research Laboratory, AMRL., TR. 78-2, 1978.

Lishman, J.R. and Lee, D.N., The autonomy of visual kinaesthesis. Perception, 2, 287-294, 1973.

Longuet-Higgins, H.C. and Prazdny, K., The interpretation of a moving retinal image. Proc. R. Soc. Lond. B 208, 385-397, 1980.

Löwenstein, O. and Roberts, T.D.M., The equilibrium function of the otolith organs of the thornback ray (Raja Clavata), J. Physiol., Vol. 110, 392-415, 1949.

Löwenstein, O. and Saunders, R.D., Otolith-controlled response from first-order neurons of the labyrinth of the bull frog (Rana

catesbeiana) to changes in linear acceleration. Proc. R. Soc.
    Lond. B191, 475–505, 1975.
McCabe, B.F., Nystagmus response of the otolith organs. Laryngoscope,
    74, 372–381, 1964.
Mach, E., Grundlinien der Lehre von den Bewegungsempfindungen. Verlag
    von Wilhelm Englemann, 1875.
Malcolm, R. and Melvill Jones, G., Erroneous Perception of Vertical
    Motion by Humans Seated in the Upright Position. Acta Oto-
    largyn, Vol. 77, 274–283, 1974.
Mayne, R., A system concept of the vestibular organs. In: H.H.
    Kornhüber (Ed.). Handbook of Sensory Physiology. Vol. VI/2,
    Vestibular system. Springer-Verlag, Berlin, Heidelberg, New
    York, 1974.
Mayne, R. and Belanger, F., The interpretation of single or few
    fiber recordings, GERA-1113, 18 pp, 1966.
Meiry, J.L., The vestibular system and human dynamic space orienta-
    tion. Sc.D. Thesis, MIT, Cambridge, Mass., 1965.
Melvill Jones, G. and Milsum, J.H., Neural response of the vestibu-
    lar system to translational acceleration. In: Supplement to
    Conference on Systems Analysis. Approach to Neurophysiological
    Problems, Brainerd, Minn., 8–20, 1969.
Melvill Jones, G. and Young, L.R., Subjective detection of vertical
    acceleration: a velocity dependent response? AMRU Reports
    Vol. V, 245–255, 1976.
Melvill Jones, G., Rolph, R. and Downing, G.H., Comparison of human
    subjective and oculomotor responses to sinusoidal vertical
    linear acceleration. Acta Oto-Laryngologica, in press (1980).
Money, K.E. and Scott, J.W., Functions of separate sensory receptors
    of non-auditory labyrinth of the cat. Amer. J. Physiol., 202/6,
    1211–1220, 1962.
Niven, J.I., Hixson, W.C. and Correia, M.J., Elicitation of horizon-
    tal nystagmus by periodic linear acceleration. Acta Oto-Laryng.,
    (Stockh.), 62, 429–441, 1966.
Ormsby, C.C., Model of Human Dynamic Orientation. Ph.D. Thesis, MIT,
    Cambridge, Mass., January 1974.
Pavard, B. and Berthoz, A., Linear acceleration modifies the per-
    ceived velocity of a moving visual scene. Perception, 6,
    529–540, 1977.
Raphan, T., Cohen, B. and Matsuo, V., A velocity storage mechanism
    responsible for optokinetic nystagmus (OKN) optokinetic after
    nystagmus (OKAN) and vestibular nystagmus. In: R. Baker and
    A. Berthoz (Eds.). Control of Gaze by Brainstem Neurons.
    Elsevier, Amsterdam, 37–48, 1977.
Robinson, D.A., Vestibular and optokinetic symbiosis: an example of
    explaining by modelling. In: R. Baker and A. Berthoz (Eds.).
    Control of Gaze by Brainstem Neurons. Elsevier, Amsterdam,
    49–58, 1977.
Salvatore, S., Velocity sensing. Highway Research., Record 282,
    79–90, 1968.
Schmid, R., Zambarbieri, D. and Sardi, R., A mathematical model of

the optokinetic reflex. Biol. Cybern, 34, 215-225, 1979.

Schmidt, F. and Tiffin, J., Distortion of drivers' estimates of automobile speed as a function of speed adaptation. J. Appl. Psychol., 53, 536-539, 1969.

Schöne, H., Orientierung in Raum. Wissenschaftliche Verlagsgesellschaft mbH Stuttgart, 377pp, 1980.

Schor, R.H., Response of cat vestibular neurons to sinusoidal roll tilt. Exp. Brain Res., 20, 347-362, 1974.

Steer, R.W., Jr., The Influence of Angular and Linear Acceleration and Thermal Stimulation on the Human Semicircular Canal. Sc.D. Thesis, MIT, Cambridge, Mass., 1967.

Stein, St. von, Schwindel (Autokinesis externa et interna). Leipzig: O. Lessier, 1910.

Travis, R.C. and Dodge, R., Experimental analysis of the sensorimotor consequences of passive oscillation, (rotary and rectilinear). Psych. Mon., 38, 1-96, 1928.

Tokunaga, O., The influence of linear acceleration on optokinetic nystagmus in human subjects. Acta Otolaryngol, (Stockh.), 84, 338-343, 1977.

Urbantschitsch, V., Uber Störungen des Gleichgewichtes und Scheinbewegungen. Z. Ohrenheilk., 31, 234-294, 1897.

Vidal, J., Jeannerod, M., Lifschitz, W., Levitan, A., Rosenberg, J. and Segundo, J.P., Static and dynamic properties of gravitysensitive receptors in the cat vestibular system. Kybernetik, 6 pp, 1971.

Walsh, E.G., The role of the vestibular apparatus in the perception of motion on a parallel swing. J. Physiol., 155, 506-513, 1961.

Walsh, E.G., The perception of rhythmically repeated linear motion in the horizontal plane. Br. J. Psychol., 53, 439-445, 1962.

Walsh, E.G., The perception of rhythmically repeated linear motion in the vertical plane. Quart. J. Exp. Physiol., 49, 58-65, 1964.

Wilson, V.J. and Melvill Jones, G., Mammalian Vestibular Physiology. Plenum Press, New York, 1979.

Wist, E.R., Diener, H.C., Dichgans, J. and Brandt, Th. Perceived distance and the perceived speed of self-motion: linear versus angular velocity. Percept. Psychophys., 17, 549-554, 1975.

Young, L.R., Effects of linear acceleration on vestibular nystagmus. Third symposium on the role of the vestibular organs in space exploration: NASA SP-152, 1967.

Young, L.R., Role of vestibular system in posture and movement. In: V.B. Mountcastle (Ed.), Mosby, St. Louis, Vol. 1, 1974.

Young, L.R. and Meiry, J.L., A revised dynamic otolith model. Third symposium on the role of the vestibular organ in space exploration, 1967. NASA SP-152, 1968.

Young, L.R., Meiry, J.L. and Li, Y.T., Control engineering approaches to human dynamic spatial orientation. Second Symposium on the Role of the Vestibular Organs in Space Exploration. NASA SP-115, held at Ames Research Center, Moffett Field, CA, 217-229, January, 25-27, 1966.

Zacharias, G.L., Motion sensation dependence on visual and vestibular cues. Ph.D. Thesis, MIT, Cambridge, Mass., 1977.

NEURAL SUBSTRATES OF THE VISUAL PERCEPTION OF MOVEMENT

Mark A. Berkley

Department of Psychology
Florida State University
Talahassee, Florida

INTRODUCTION

The purpose of this paper is to provide a brief summary of recent ideas and data regarding the neural substrates of the perception of motion. This summary is not comprehensive nor does it include the considerable relevant literature in human psychophysics. Several recent detailed and outstanding reviews are available also covering movement perception; in particular, the reviews by Grüsser and Grüsser-Cornellis (1975) and by Sekuler, Pantle and Levinson (1978) are notable. These reviews attempt to take into account the many physiological findings that may be relevant to motion perception. Several earlier, less physiological reviews are also available in basic reference texts and handbooks, e.g., Walls (1942), Johansson (1970), Graham (1963), etc., and the reader is referred to these for general background.

Even the most casual consideration of vision would quickly convince anyone that visual stimulus motion (image motion) is a ubiquitous component of our everyday visual experience. It is not surprising, therefore, that interest in this dimension of vision has fascinated so many investigators (see reviews mentioned above). Since this summary is primarily concerned with data derived from animals (rather than humans), the work of Walls (1942) is of particular interest because he attempted to relate the ability to perceive motion to certain aspects of life-style and motor capabilities of different animals. We shall return to this point later on in this paper.

1. DEFINITIONS

Before proceeding further, it is important to delineate what, in the present context, stimulus conditions are the necessary antecedent stimulus conditions for motion perception and to provide several definitions.

1.1. Real Movement

In all cases the adequate stimulus for producing a perception of motion is the movement, or displacement of an image on the retina. This image displacement can be produced by one of several mechanisms. For example, (1) the movement of a real object on the retina; (2) movement of the eyes with the objects in space being stationary (or moving) will also produce image movement on the retina and (3) the displacement or movement of several points or contours, relative to each other, can produce a perception of an object (Johannson, 1970). Several special conditions are also important in movement perception, e.g., the relative movement (object against a stationary background or relative to other objects in the field) and movement of the entire visual field (streaming). To describe such stimuli, the dimensions of speed (degrees of visual angle/sec) and directions of movement (meridian) are usually used. The latter dimension usually being specified in polar coordinates i.e., 0° - 360°.

1.2. Apparent Movement

In addition to real image movement on the retina, whether it is induced by real image movement or whether it is induced by eye movement, a perception of movement can be achieved by a sequential displacement of an image without real continuous movement of the object having taken place, apparent motion or the so-called "phi" phenomenon. This phenomenon points out the importance of the temporal characteristics of the visual system. While such a stimulus can be shown to mimic real movement in many ways and has been the subject of intense research interest, it is a special stimulus condition and will not receive any attention in this paper except to note that some physiological measures have been made using this type of stimulation.

1.3. Temporal Response

While apparent movement will not be discussed here, it does serve to point out that time (e.g., latency, persistence) is an important parameter of motion perception and it has long been recognized as being critical in understanding motion perception. Thus, the amount of time it takes either to process the image or produce the transduction of light into signals will influence the limit of movement perception. One measure used to estimate the temporal time

course of visual mechanisms, and its relation to movement sensitivity, is flicker sensitivity, and more recently, spatio-temporal modulation sensitivity. These measures will not be discussed in detail
but will be considered in the review of measures of possible neural
substrates.

1.4. Stimulus Dimensions

In considering the possible neural mechanisms underlying motion
perception, several sensitivity procedures derived from psychophysics have been used and will be referred to in considering physiological data. Thus, the visual system has been characterized using:
(1) spatial and temporal frequency analysis (spatial and temporal
contrast sensitivity), and (2) directional selectivity (e.g.,
Campbell & Robson, 1968; de Lange, 1954).

In addition, several special stimulus dimensions have also
been examined in the search for neural substrates, e.g., movement
parallax, looming, relative image motion and streaming, less quantitative psychophysical data is available for these dimensions.
While not exhaustive, these dimensions represent the major parameters of movement vision that have been studied physiologically
and can be related, in one way or another, to movement perception.

2. CRITERIA FOR ESTABLISHING THE NEURAL SUBSTRATES OF MOVEMENT PERCEPTION

The number of procedures available for discovering neural substrates are rather limited and have been derived from the general
approach of neurobehavioral analysis. These methods fall into two
broad categories: (1) direct measurement methods and, (2) indirect
methods.

2.1. Direct Methods

1) Correlational methods. One of the main techniques employed
in the search for neural substrates is to compare physiological and
psychophysical measures of a particular stimulus dimension in the
same animal. Usually, the physiological measures are made in an-
esthetized animals, but to more accurately simulate natural conditions, they can be made in awake, moving animals (e.g. Bridgeman,
1973a and b). There are several difficulties in this kind of approach. One is the difficulty in obtaining both classes of data
from the same organism. In measurement of movement sensitivity, the
problem of distinguishing the response to real stimulus movement
from self-induced retinal movements (Wurtz, 1969; Goldberg & Wurtz,
1972a, b) is also difficult. We will return to this point later.

2) Manipulation of putative neural substrate. Another direct
method by which one can estimate the contribution of a particular
neural population to the perception of movement is to manipulate

the suspected neural population and note the effect on the perception of movement. One such procedure is to remove or ablate a
population of neurons which is thought to participate in movement
perception (Anderson & Symmes, 1969; Kennedy, 1936, 1939; Smith,
1941).

A second manipulation method is derived from recent experiments
on the development of the visual system. In this case, one can produce changes in "selected" cell populations that may participate in
movement by manipulating the visual environment during the development of the organism (Daw & Wyatt, 1974, 1976; Riesen & Aarons,
1959). Specifically, the procedures of monocular deprivation and
strobe-rearing have been shown to produce changes in the normal
properties of cortical neurons and their response to movement (although there are probably other neural changes as well, even though
they have not been investigated). A third deprivation procedure
exposes young animals to only one direction of movement of the visual
scene. All of these procedures have been shown to have measurable
physiological effects on many movement sensitive neurons (Cynader
et al., 1975; Cynader & Chernenko, 1976; Pasternak, personal communication), and thus have provided an opportunity to correlate the
noted physiological effects with behavioral estimates of movement
perception.

2.2. Indirect Measurement Methods

The methods described above require direct intervention either
through recording procedures, ablation procedures or deprivation
methods. A second stratagem for attempting to find neural substrates employs indirect measurement methods. Basically, such
methods are also correlational in that a behaviorally measured
capacity is compared to related electro-physiological measures
(Berkley et al., 1978b). In this instance, however, the measurements
are not necessarily made in the same animal or with parametric
(quantitative) techniques. In some cases, one can take advantage
of unusual features of the perception of movement in these comparisons. For example, prolonged viewing of a moving scene produces an
after-effect and, thus, post-stimulation effects in neural populations. Such observations have been compared with behavior observations (e.g., Vautin & Berkley, 1979).

Since it is not ordinarily possible to make physiological or
anatomical measurement on human subjects for comparison with behavioral measures, in using this stratagem one has to rely primarily
on comparisons of human psychophysics and animal neurophysiology
and anatomy (e.g., Kennedy & Smith, 1935; Beverly & Regan, 1973;
Camisa, et al., 1977; Richards & Smith, 1969; Richards, 1971). It
has been pointed out many times that there are pitfalls in this
kind of comparison because: (1) the neural-visual system in nonhuman primates differs from that in the human, making comparisons
somewhat tenuous; and (2) often it is not clear whether the animal
model has the same visual capacity as that of humans despite appar-

rent nervous system similarities. Consequently, it is best to attempt to make both classes of measurements in animal models, that is, to employ direct methods such as the correlation of psychophysical and electrophysiological measures in same animal models. Despite the advantages of this tactic in looking for neural substrates, it has not been widely employed, probably because of the difficulty in doing the necessary animal psychophysics. Animal psychophysics, however, is a vital link in all the methods described as it provides the information necessary to compare with electrophysiology and with human psychophysics.

3. EXPERIMENTAL DATA

3.1. Neurons that Respond to Moving Stimuli

In the previous section, the tactics and stratagems that have been employed in attempts to discover the neural substrates of vision, and of movement perception in particular, have been reviewed. In the following sections, I will take specific examples of experimental data which demonstrate an application of the procedures mentioned above with regard to the neural substrates of movement perception. As a beginning, the electrophysiological data showing the response of neurons to moving stimuli will be briefly reviewed. While the fact that a neuron responds to moving stimuli does not mean it participates in movement perception, for the purposes of this presentation it is assumed that they may. This section will be divided into the various locales in the nervous system in which such neurons have been measured.

1) The retina. While most of the studies of mammalian retina have demonstrated that the receptive fields of ganglion cells in this structure are, for the most part, concentrically organized, and not specialized for signaling movement, there have been a number of reports indicating that at least in rabbits, some neurons of the retina respond differentially to moving stimuli (Barlow & Levick, 1965; Levick, 1967; Wyatt & Daw, 1975). Experiments in the rabbit indicate that there are directionally selective ganglion cells in the retina whose movement selectivity cannot be predicted based on the map of the receptive field obtained with static stimuli (Barlow & Levick, 1965). Directional sensitivity in the retina, however, is primarily found in the rabbit among the mammalian species as cats and monkeys do not appear to possess any significant number of directionally-selective ganglion cells (Rodieck & Stone, 1965; Rodieck, 1973). (Cats may have some but they have more diffuse properties than those seen in rabbits.)
Ganglion cells of the retina have also been classified on the basis of their spatio-temporal response characteristics (Enroth-Cugell & Robson, 1966). This classification scheme, originally pro-

posing 2 classes of cells (Enroth–Cugell & Robson, 1966) has recently
been expanded to subdivide ganglion cells in one of 3 classes, X, Y
or W (Rowe & Stone, 1977). The details of this scheme are beyond the
scope of this paper but suffice it to say that the properties of
Y–cells (e.g., rapid conduction velocity, highest temporal response,
relatively poor spatial tuning) make them attractive candidates as
the first order input to the movement detection system. More will
be said about this classification scheme in later sections.

      2) Subcortical structures. The study of motion sensitive neu-
rons has been extended to numerous other structures such as lateral
geniculate nucleus and the superior colliculus as well as the
cerebellum and pontine nuclei. Except for the rabbit, essentially
all the neurons of the lateral geniculate of mammalian species that
have been studies to date appear to have concentrically organized
receptive fiels (Kozak et al., 1965; Levick, Oyster & Takahashi,
1969) or show non-specific responses to stimulus movement. Neurons
in LGN respond to moving contours but, except for showing some minor
asymmetries, are not directionally selective (Kozak et al., 1965).
Thus, this nucleus appears to be a poor candidate for the site of
movement perception. The rabbit, however, shows considerable
sharpening of directional selectivity in the LGN (Levick et al.,
1969).
      There is some suggestion that neurons in LGN may signal
stimulus velocity in that some cells show an increase in peak fir-

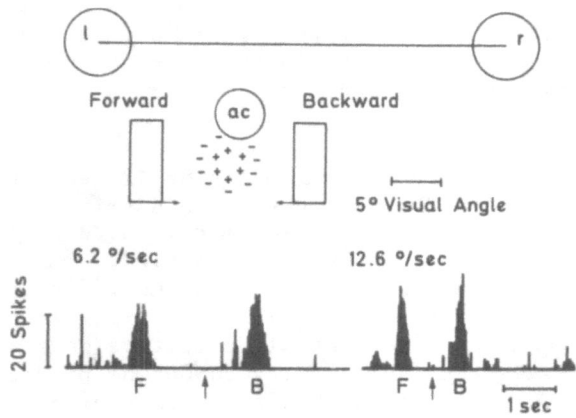

Fig. 1. Response of single cell in LGN to movement of a bar stimulus
        through the receptive field. Top portion shows receptive
        field and stimulus configuration. Bottom portion shows the
        peristimulus histograms of neurons' response. F indicates
        movement of stimulus in one direction; B indicates stimulus
        movement in the opposite direction. Arrow indicates point
        (in time) of reversal of direction of stimulus movement.
        From Hess & Wolters (1979).

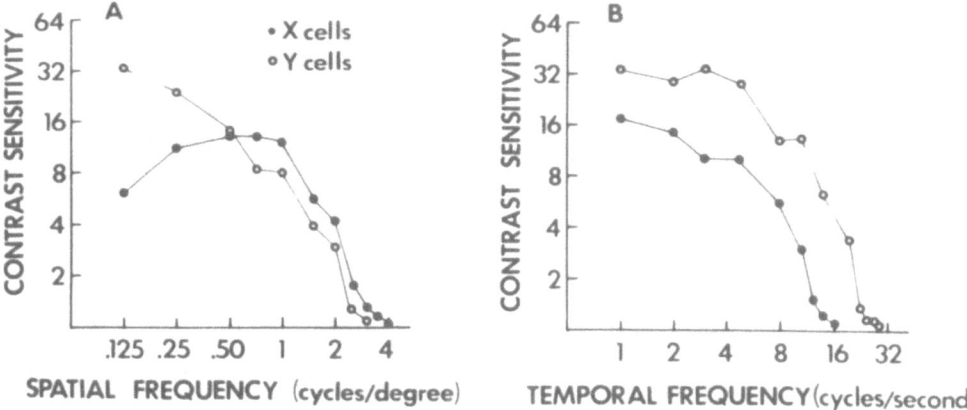

Fig. 2. Spatial and temporal contrast sensitivity of LGN cells
estimated from electrophysiological measures for two
"classes" of cells in LGN. (From Lehmkuhle et al., 1980.)

ing rate with increases in stimulus velocity. An example of this
response is shown in Fig. 1 adapted from Hess & Wolters (1979).

The subdivision of neurons by their functional properties
mentioned earlier (X-Y-W) has been carried into the LGN. The segregation of neurons seen in the retina appears to be maintained in
this nucleus, e.g., neurons have been classified X, Y and W based
on criteria developed for classifying retinal ganglion cells.

These various subdivisions may relate to movement perception,
and their response characteristics have been described using the
same linear systems analysis methods that have been used in human
psychophysics (contrast sensitivity functions). Fig. 2 shows a set
of sensitivity functions derived from the responses of neurons in
LGN showing the differences in spatial and temporal sensitivity of
X and Y type cells. Human psychophysical studies have suggested a
similar dissociation between the mechanisms for detection of moving
versus stationary gratings, although this simple view has recently
been challenged (Lennie, 1980).

In the cat superior colliculus, many of the cells in the super-
ficial layers have been found to be directionally selective and
responsive to movement. Sterling & Wickelgren (1969) showed that
most collicular cells responded vigorously to moving stimuli and
were directionally selective. The majority of these cells seem to
prefer directions which were parallel to the horizontal meridian.
Much of the directional selectivity of these cells, however, appears
to be derived from an input from visual cortex because when visual
cortex is either cooled or removed many of the directionally selective cells in the superior colliculus lose their selectivity although
they may still respond to movement (Wickelgren & Sterling, 1969;
Sterling & Wickelgren, 1969, 1970). Directional selectivity in the

monkey colliculus, is, however, less dependent on cortex (Schiller et al., 1974).

We shall return to the discussion of movement selective cells in the superior colliculus in a later section where self-induced motion is considered.

Other neurons that have been shown to respond to stimulus movement have been found in the medial pontine nuclei of the cat (Glickstein et al., 1971, 1972; Glickstein & Gibson, 1976). This region, which has been shown to receive an input from extra-striate cortex (Gibson et al., 1978), and different from the pontine region receiving an input from superior colliculus, responds primarily to

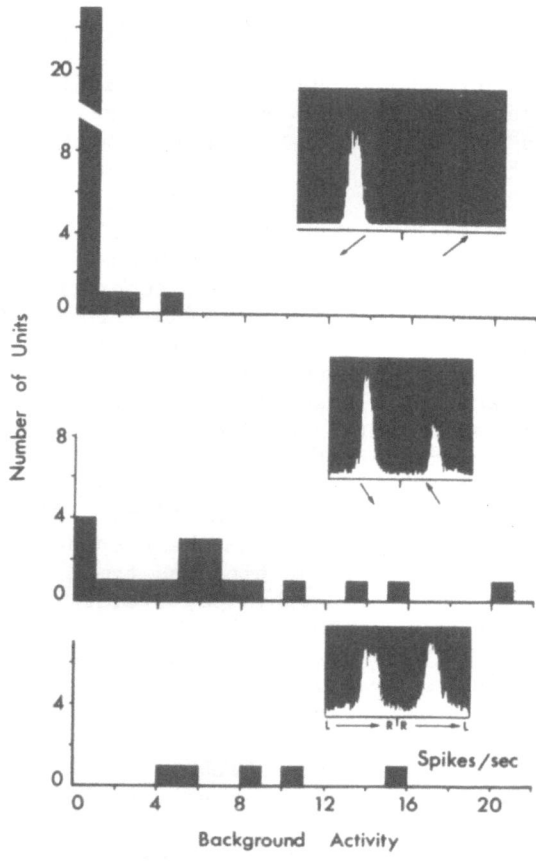

Fig. 3. Representative distribution of the number of cells in cat visual cortex showing differing degrees of directional selectivity. Insets depict prototypical peristimulus response histograms for each of the three general categories. Upper: unidirectional cells; middle: weakly bidirectional; lower: bidirectional cells. From Pettigrew et al. (1968).

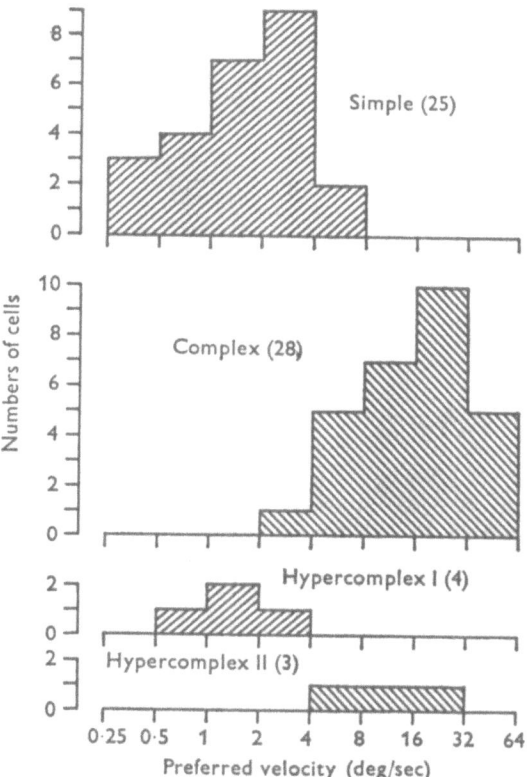

Fig. 4. Histograms depicting differences in stimulus velocity
        preferences in different classes of cortical neurons of
        the cat. From Movshon (1975).

large textured fields moving in a specific direction. Receptive
fields are very large, do not respond well to small moving stimuli,
and have poorly defined borders (Baker et al., 1976). They appear
to be ideally situated to signal movement of the visual scene to
the cerebellum and thus to provide sensory information to the
muscular control system (Buchtel et al., 1973).

    3) Neocortex. Many studies have shown cells in the neocortex
to be sensitive to stimulus movement beginning with the studies of
Hubel and Wiesel (1962) and Baumgartner et al. (1964). Other
studies have subsequently shown that some striate cortical neurons
are specifically selective for direction of movement (see Fig. 3;
Pettigrew et al., 1968) and to some extent, velocity (see Fig. 4;
Movshon, 1974, 1975).
    For many theorists who believe that perception is a cortical
function, these directionally selective movement cells have been

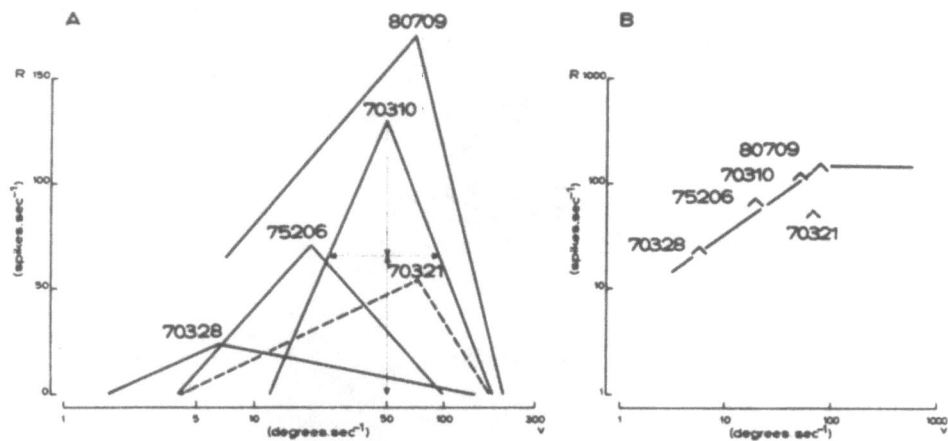

Fig. 5. A. Velocity tuning in cells in area 18 of the cat.
B. Plot of apices of velocity tuned cells plotted with the
mean response of velocity sensitive cells (continuous lines).
From Orban & Callens, 1977.

assumed to be the neural substrate of movement perception. This
belief is strengthened by the fact that many·properties of these
movement selective cells are similar to the psychophysical dimensions of movement perception, e.g., they are directionally selective
and they are broadly tuned for velocity (Movshon, 1975). It is well
to keep in mind, however, that most of these studies have been performed on cats, an animal whose central visual pathways are organized differently from primates (Berkley, 1976).

From these and many similar studies, it is clear that contour
movement is a ubiquitous and powerful activator of neurons in striate
cortex (e.g., Creutzfeldt et al., 1974; Hammond, 1978). Because
cortical neurons respond so poorly to stationary stimuli and appear
to be tuned to respond to stimulus change, the response to movement
is hardly surprising. Do all neurons in striate cortex signal movement or only some specific subpopulation? It is possible to subdivide the cortical cell populations into groups that are (1) directionally selective, (2) velocity tuned, or (3) respond to texture
movement even though they all reside in the same cortical area.
Recent studies (e.g. Orban and Callens, 1977) have shown that movement sensitive cells could be subdivided into various categories,
based on their responses to different stimulus velocities (see
Fig. 5).

These results suggest a possible segregation of movement -
signalling neurons from those concerned with contour analysis.
Whether such a separation of function exists in the primate is still

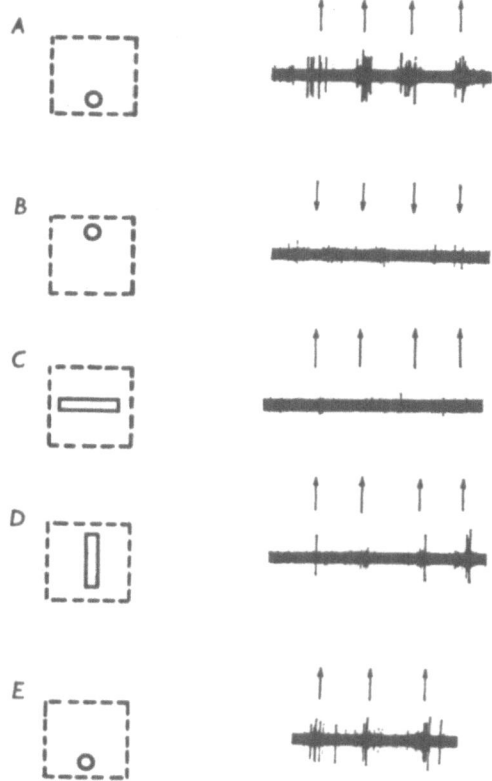

Fig. 6. The response of a cell in superior temporal sulcus of the
rhesus monkey to moving spots and bars. Stimulus configuration and receptive field shown on left; action potentials
produced by stimulation in direction depicted by arrows
shown on the right. From Zeki (1974a).

unclear. Recently, Zeki (1974) has found a region outside of
striate cortex where the receptive fields are very large, respond
to movement anywhere in the field, are directionally selective and
are relatively insensitive to the spatial form of the visual stimulation. He suggests that this zone may be specially adapted for the
perception of movement (Zeki, 1974a) (see Fig. 6). While these
findings are highly suggestive, they have yet to be validated using
behavioral methods.

Based on the work of Orban and colleagues in the cat and Zeki
in the monkey, it is tempting to suggest that the neural mechanisms
involved in movement perception are segregated from other visual
neural mechanisms. It should be pointed out, however, that the
similarity between the cat and monkey results may be more apparent
than real because of the differences in the anatomical organization
of the extrastriate system in both animals. Secondly, while the
physiological data seems to suggest a functional segregation, the
idea has not been subjected to empirical tests using behavioral
methods.

3.2. Neurons that Respond to Special Classes of Movement

In the previous section, I briefly outlined the various
locales in the visual nervous system where neurons that respond to
movement have been found. In this section, I will describe the
results of the experiments that were specifically looking for cells
responding to unique dimensions of stimulus movement, e.g., response
to differential rates of movement in each eye, response to looming
stimuli, the habituation of neurons to repeated stimulation with
moving stimuli etc.

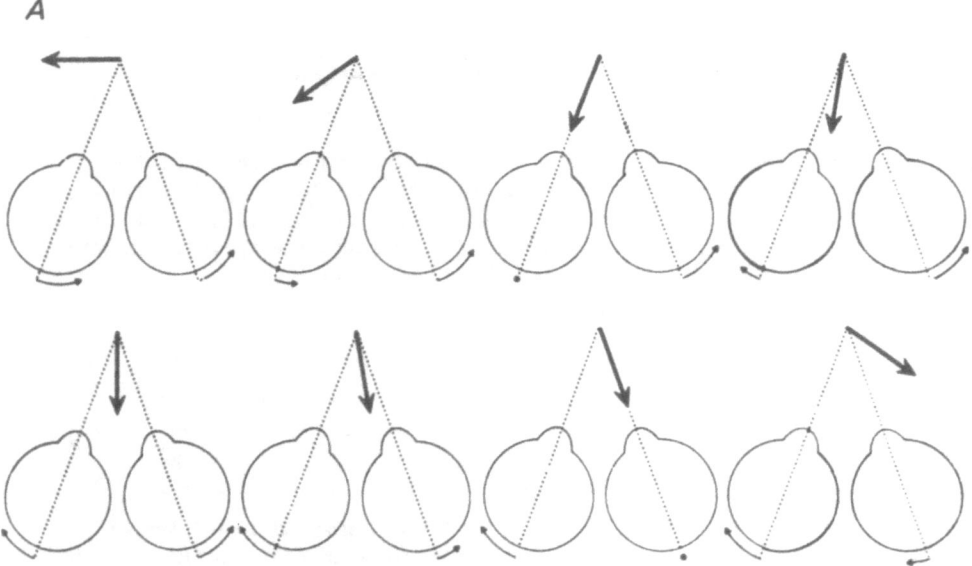

Fig. 7. Schematic depiction of different velocities of movement
        produced on each retina by objects moving in different
        directions in 3 dimensional space to demonstrate the
        stimulus conditions that would optimally stimulate some
        cells in area 18 of the cat. From Cynader & Regan (1978).

1) Movement parallax and looming. In recent studies, Cynader & Regan (1978) have found neurons responding to differential rates of movement in each eye in the cat. On the basis of geometric optical considerations, they concluded that these cells were uniquely selective for responding to movement parallax or movement of objects in depth. Thus, an object whose trajectory in space is not parallel or orthogonal to the frontal plane but oblique to this plane will produce image movement on each retina at different velocities. Therefore, those cells that are selective for different-ial movement rates in each eye could signal the movement of objects in such planes (see Fig. 7).

Obviously, movement of objects in the fronto-parallel plane, in any meridian, produces in each eye or within a receptive field an equal movement of all stimulus contours. That is, the speed and direction of the contour movement is the same in each eye or at all edges of the stimulus. For objects (visual stimuli) which are moving toward or away from the eye, this condition is not true, that is, the edges of the contour are moving in opposite directions. Such stimuli would be described as moving in a direc-tion approaching the eye if the edges are moving temporally (loom-ing) and receding if they were moving nasally. Cells whose lateral edges of the receptive field prefer movement in opposite directions are candidates for such a detector in that they could signal the presence of looming or receding objects. Such cells have been de-scribed by Zeki (1974b) in the monkey (see Fig. 8) and by Regan and Cynader (1979) in extrastriate cortex of the cat.

2) Habituation of movement selective neurons. Based upon perceptual studies, the mechanisms processing movement are known to be affected by prior stimulation. For example, the waterfall illusion (movement aftereffect) is presumably due to the after-effects of repeated stimulation of that neural population which normally responds to moving stimuli (Sutherland, 1961; Barlow & Levick, 1965; Sekuler et al., 1978). Insofar as it is possible to demonstrate a neural population which shows after-effects of the same magnitude and characteristics as the perceptual after-effects, one could conclude that cells that have this property at least have the capability of contributing to movement perception. Two physio-logical studies of the after-effects of movement stimulation have been reported, one by Maffei et al. (1973) and the other by Vautin and Berkley (1977). In both cases, cells in the striate cortex of the cat were shown to habituate to repetitive stimulation with moving stimuli (see Fig. 9).

It was also shown that the habituation was directionally selective and was not due to peripheral adaptation, that is retinal adaptation, and had a time course similar to that observed in human studies of the movement after-effect (see Fig. 10 and 11). Response habituation appears to be a ubiquitous property of neurons in the striate cortex (although Maffei et al. (1973) reported that

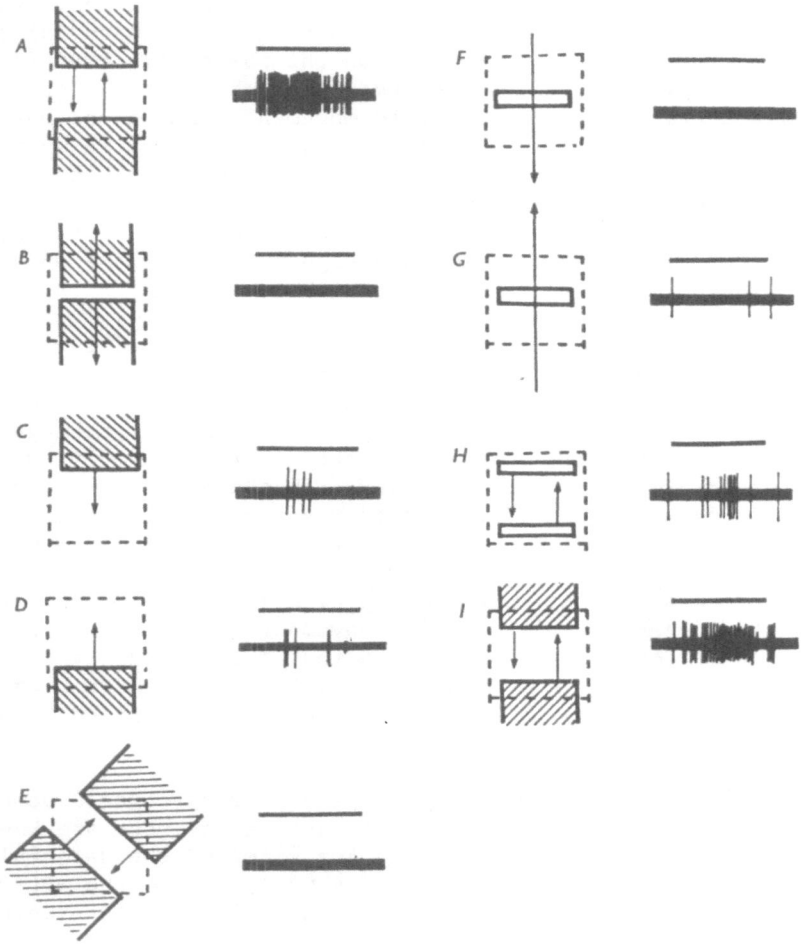

Fig. 8. Response of binocularly activated cell in superior temporal
sulcus of the rhesus monkey to edges or bars moving toward
or away from each other as well as control data showing
response to single bars and edges or non-optimally oriented
edges. Stimulus configuration and receptive fields shown to
the left of each sample response. Duration of sweep about
4 sec. From Zeki (1974b).

only complex cells habituate). According to the report by Vautin
and Berkley (1977) all cells that were orientation selective and
directionally selective showed the characteristics of habituation.
It is not clear from these studies, however, if there is a subpopulation which participates specially in perception of movement.

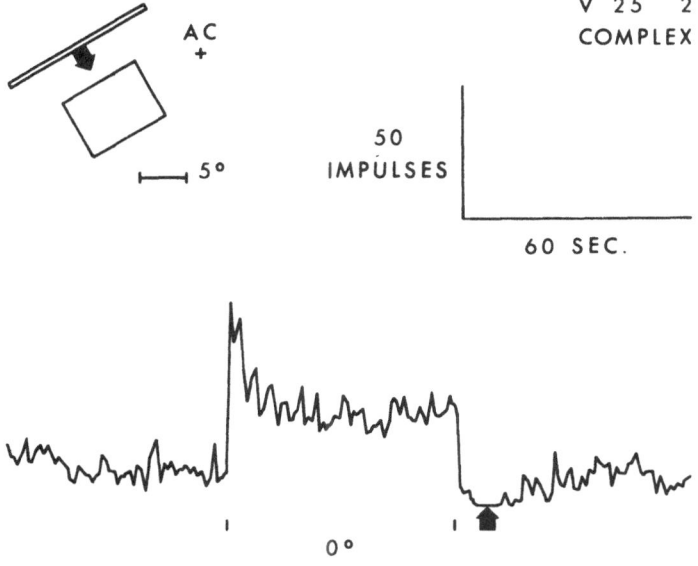

Fig. 9. Response of a complex cell in cat cortex to repeated
stimulation with a preferred stimulus. Top left: receptive
field size, locus and stimulus are depicted. Bottom: first
segment shows 1 min of spontaneous activity; middle section shows response to 1 min of stimulation with a preferred stimulus; last segment shows activity of cell for
1 min after cessation of stimulation. Note silent period
at arrow. From Vautin and Berkley (1977).

    3) Whole field movement and streaming. Several investigators
have demonstrated that there are neurons in the visual cortex that
respond best to the movement of large (whole) fields (Hammond and
MacKay, 1975). The receptive fields of these cells are large, and
may prefer a specific direction of movement. They would be activated when the animal was moving (Hammond and MacKay, 1977, 1978;
Regan & Beverly, 1979). These cells are different from the small
field, movement sensitive cells and may be more concerned with
locomotion than movement perception.

3.3. <u>Neurons that Respond Differentially to Self-Induced Movement</u>

    In discussing neuron populations which have been shown to
respond to movement of visual contours, one cannot ascertain whether

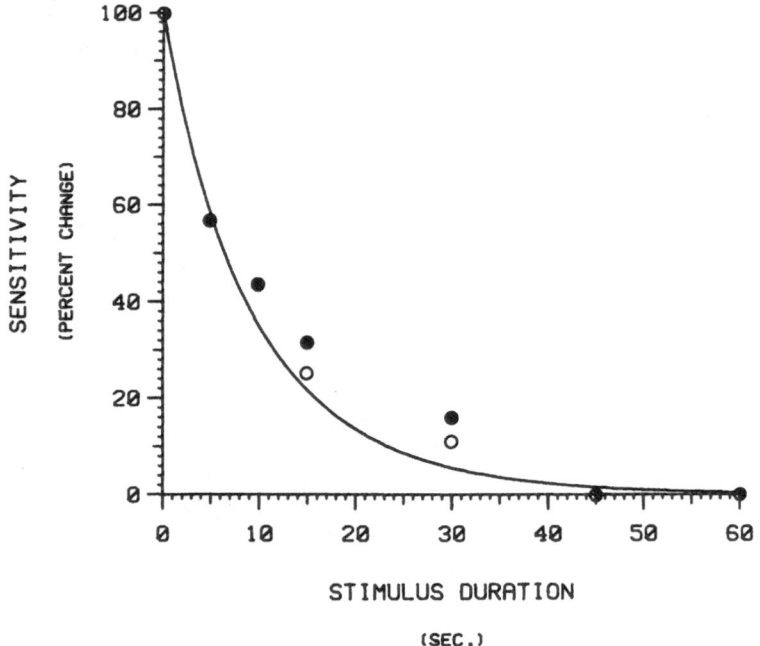

Fig. 10. Graph showing mean loss in responsiveness of cortical
neurons in area 17 as a function of duration of stimulation with moving bars or gratings (continuous line). Open
and filled circles show change in contrast sensitivity
of human observers after adaptation to moving gratings
for different durations as described by Blakemore and
Campbell (1969). From Vautin and Berkley (1977).

these cells would respond to movement of contours if the contour
movement on the retina were generated by a head or eye movement
rather than object movement. It is clear that the nervous system
is capable of distinguishing between self-induced retinal image
motion from real image motion. This aspect of neural signalling
has been the subject of a considerable amount of investigation
(e.g., Monty and Senders, 1976), The details of these investigations, however, are beyond the scope of the present presentation
and the reader is directed to any recent book on eye movements.
However, it is important to point out that a number of studies
which have looked at movement-selective cells have specifically
addressed the issue of whether or not these cells respond differ-

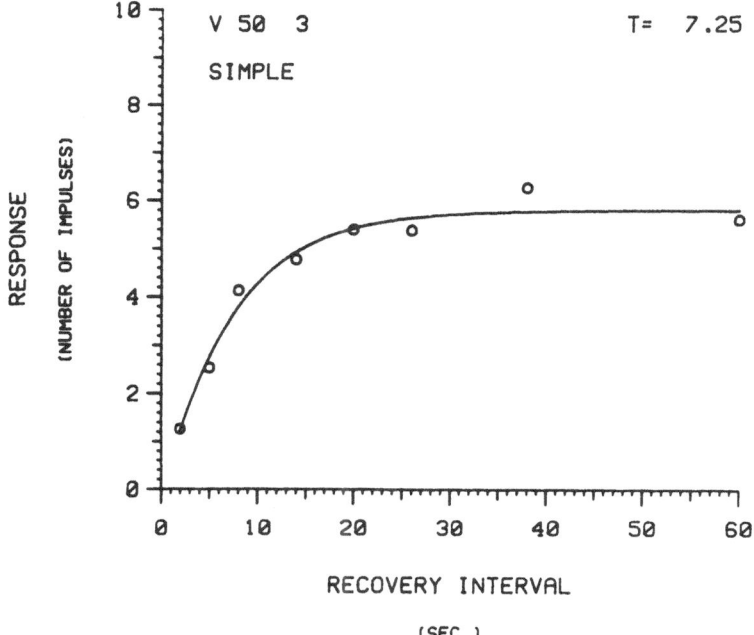

Fig. 11. Time course of recovery of responsiveness of a simple cell
in area 17 of the cat after it had been stimulated continuously for 1 min. From Vautin and Berkley (1977).

entially to retinal self-induced image motion and real image motion
(e.g., Schiller and Koerner, 1971; Bridgeman, 1972; Wurtz, 1969;
Mays and Sparks, 1980). In the studies of the retina of most mammals,
there are relatively few ganglion cells that respond selectively
to movement. Thus, the retina is not a good candidate as the site
of movement perception. In addition, there is little, if any,
evidence that suggests a feedback pathway into the retina of mammals
that could provide a signal (corollary discharge) that could
differentially signal self-induced image motion (Rodieck, 1973).

In the lateral geniculate nucleus, similar arguments could be
made, e.g., cells here are not directionally selective (but see
Montero and Brugge, 1968). In the geniculate, however, there may
be a neural substrate which is capable of providing a feedback
signal indicating that the image motion that is being signalled
by the retina is due to self-induced movement. There is evidence
for inputs to the geniculate from cortex (Guillery, 1967; Kawamura

et al., 1974; Hollander, 1972) as well as from midbrain structures
(Singer and Bedworth, 1978) that could provide such information.
A recent study by Singer and Bedworth (1973), however, suggested a
somewhat simpler mechanism. These authors suggest that the different
classes of retinal ganglion cells that project to LGN (X-Y cells)
can internally provide their own feedback signal for suppression or
for cancellation of the self-induced movement. Singer and Bedworth
(1973) suggest that a rapid eye movement or change in position of
the eye can activate the Y-cell population which could then inhibit
the X-cell input to LGN, providing a signal for the suppression of·
vision during eye movement. They did not, however, address the
issue of how the nervous system then distinguished Y-cell activity
generated from a real image motion from a self-induced image motion.

     A more likely locus for the signal which is capable of separat-
ing a real from a self-induced image motion is the superior colliculu-
lus. A number of studies which have examined single units in these
areas, most notably the work of Wurtz (1969), have addressed this
issue (Wurtz, 1969; Robinson and Jarvis, 1974; Robinson and Wurtz,
1975). These authors were able to find cells in the superior colliculus which responded to visual stimulation when the eye was
stationary but not to visual stimulation when the eye was moving.
They concluded that there was an extra-retinal signal that produced
this differentiation (suppression) and that the cells which are
thus suppressed can provide the signal to distinguish between real
stimulus movement and stimulus movement resulting from the animal's
own movements. The origin of the input to the superior colliculus
cells is still unclear but probably is derived from those neural
areas originating eye movements, e.g., frontal eye field (Schiller
et al., 1979). The output of these movement differentiating cells,
is still not known. Apparently, it is not the striate cortex because
Wurtz (1969) has shown that striate cortical neurons which respond
to movement show the same response whether the image movement was
self-induced or real.

     In any case, there is sufficient evidence to suggest that there
is a unique cell population which can distinguish two kinds of
image motion. How these cells may participate in perception of movement is still, however, not understood.

3.4. Selective Ablation Studies

     1) Superior colliculus. Despite the fact that neurons that
respond to retinal image motion can be found in a variety of
structures, a number of investigators have attempted to examine the
ability of animals to discriminate image motion after damage or
removal of one or another of these cell populations (Smith, 1941;
Kennedy, 1936, 1939; Berkley and Sprague, 1977). For example,
animals with superior colliculus ablated have been studied for
their ability to discriminate image motion after the ablation.
Anderson and Symmes (1969) have shown that after collicular abla-

tions, cats and monkeys are able to discriminate image motion but show some impairment in their ability to distinguish different rates of movement from each other. Other studies by Berkley (unpublished observations), have also shown that the thresholds for the detecting slow movement are unaffected in cats after removal of the superior colliculus. Loss of SC in humans, while not destroying movement perception, apparently produces some difficulty in distinguishing real from self-induced image motion (Heywood and Ratcliff, 1975), a finding consistent with the single unit studies cited earlier.

2) Cortical lesions. In humans and animals with visual cortical ablations, the ability to detect motion appears to be intact in most cases although there is very little quantitative psychophysical data on these patients (Poppel et al., 1973; Koerner and Teuber, 1973; Denny-Brown and Chambers, 1976). Riddoch (1917) reported that humans with occipital pole damage (striate cortex plus some adjacent zones) are able to detect motion despite the fact that they are completely unable to detect shapes. Similar findings have been reported in monkeys (Humphrey and Weiskrantz, 1967) and also confirmed in humans (Weiskrantz et al., 1974). In one quantitative study, Berkley and Sprague (1977) found that cats without areas 17 and 18, while showing some minor deficits, had normal thresholds for slow movement (see Fig. 12). Similar data were reported by Kennedy (1939) and Smith (1941) with much larger ablations. Hamilton and Lund, (1970), using a split-brain monkey preparation, have claimed that cortex is essential for movement perception, but Peck et al. (1979), using cats in a similar paradigm demonstrated that cortex is not essential.
Thus, animals with ablations of either superior colliculus or striate cortex retain the ability to detect motion. No doubt more detailed psychophysical studies will reveal subtle losses. If one considers that the pathway from the retina that is most likely carrying information about image movement (Y-cells) projects to superior colliculus as well as cortex (Stone and Dreher, 1973), it is not surprising that ablation of either area alone does not eliminate the ability to detect motion.

3) Selective deprivation. The discovery that early visual experience can affect the properties of visual neurons has recently been used to study the neural substrates of movement. In these studies some type of visual deprivation is employed in an attempt to change the response properties of cortical neurons. The deprivation procedures may be general, e.g., binocular or monocular visual deprivation via lid suture, or selective, e.g., exposure to only one dimension of the visual scene. In the case of selective movement deprivation, neonatal animals might be permitted to view only moving visual scenes (all the objects of a scene moving in one direction and never permitted to see movement in any other direction (Daw and Wyatt, 1974, 1976; Cynader et al., 1975). Another

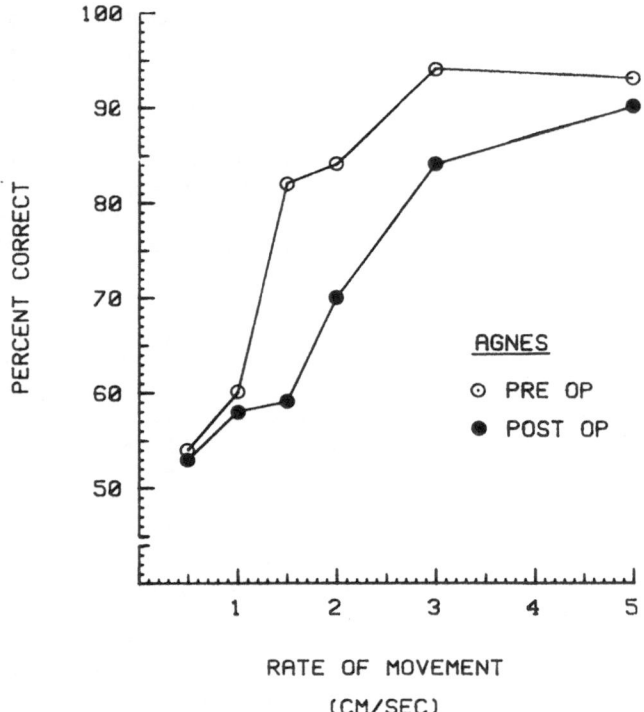

Fig. 12. Pre and post operative performance of a cat discriminating
        a moving from a stationary target as a function of moving
        target velocity expressed as the percentage of correct
        choices. Open circles indicate pre-operative performance
        while filled circles show performance after essentially all
        of cortical areas 17 and 18 have been removed. From Berkley
        and Sprague (1977).

procedure is to completely deprive animals of movement stimulation
but not contours using the technique of strobe-rearing. In this
case, animals are reared in an environment briefly and repetitively
illuminated with strobe flashes (Cynader and Chernenko, 1976). The
flashes are sufficiently short to preclude any significant stimulation with moving components of the visual scene (Tretter et al.,
1975).
     Physiological studies of these preparations (cats) show that
after monocular deprivation very few cortical cells can be activated

via the deprived eye and there are very few cells responsive to moving contours (Wiesel and Hubel, 1963, 1965). Behavioral tests of such animals indicate that they are virtually blind in the deprived eye (Dews and Wiesel, 1970; Berkley et al., 1978a; Jones et al., 1978). Even casual testing with normally visually arresting stimuli, e.g., moving strings or other small moving objects, is unable to elicit visual responses in deprived animals. Some of the early deprivation studies where animals were reared in the dark also indicated the deficits in ability to respond to moving targets (Riesen and Aarons, 1959). It is not clear from these earlier studies, however, whether the animals were impaired in their visuomotor behavior or in their ability to perceive movement. Studies by Myers and McCleary (1964), for example, have suggested that a very large component of the visual deficit observed in totally deprived animals is a result of impaired visuomotor mechanisms. However, most of the deprivation studies performed prior to 1968 employed insufficiently long deprivation periods, confounding their results with deprivation reversal effects as there is a "critical" period during which the system is susceptible to manipulation. Nevertheless, many of the effects of deprivation appear to be on the visuomotor mechanisms (Held, 1968; Vital-Durand and Jeannerod, 1974).

In the studies where unidirectional moving targets have been presented during development, physiological studies by Daw and Wyatt (1976) and Cynader et al. (1975) have shown that a majority of the movement selective cortical cells responds to movement in the exposed movement direction, while there is a significant loss or reduction in movement selective cells responding to other directions of movement. The behavioral studies of such animals are still relatively inconclusive, but the animals do respond to targets moving in the deprived directions (see Pasternak's paper presented at this symposium describing some of the behavioral results: Pasternak et al., 1981).

In the strobe-rearing studies, the animals are not exposed to any moving contours during their development. Physiological studies have shown that stroboscopic rearing is a condition sufficient to potentially produce a complete loss of directional movement sensitivity (and other changes as well) in cells of striate and extrastriate cortex of cats (Cynader and Chernenko, 1976; Orban et al., 1978) and to a great extent, in the superior colliculus as well. The behavioral testing of these animals is as yet incomplete, but preliminary reports from Pasternak and Merigan indicate that there are abnormalities in the ability to discriminate movement by these animals (as well as other abnormalities of vision) (Pasternak and Merigan, 1979; Pasternak et al., 1981.)

From these developmental studies, it is clear that manipulations that produce physiological changes in the responses of cortical cells that normally respond to moving targets are sufficient to produce gross anomalies in the behavior to moving targets by these animals. What is far from clear, however, is whether the cells that have been physiologically measured are the ones re-

sponsible for the observed behavioral deficits and thus represent
a substrate of movement perception. One difficulty with all deprivation procedures, of course, is that they are non-selective for
neural regions. Rearing procedures affect not only the neural
population that is investigated electrophysiologically, but other
neural populations as well. Thus, it is not possible to conclude
with any degree of certainty that the measured population is in
fact a population contributing to the normal movement perception.

4. CONCLUSIONS

We have seen that it is possible, via a variety of procedures,
to demonstrate correlations between the properties of neurons in
various parts of the visual system with perceptual characteristics
in movement sensation. This should come as no great surprise. Certainly there should be neurons in the nervous system which respond
to stimulus dimensions which are known to affect our perception.
What is unclear from these correlational studies is whether the
population whose behavior is correlated with perception is the substrate of that perception.

From the ablation studies a few conclusions emerge. The perception of movement is not localized in a single neural structure. A
variety of neural structures have the capability to respond to
moving image contours, but which ones normally contribute to our
perception of movement cannot be derived from studies of movement
sensitive cells. However, we can say that it is highly likely that
more than one area contributes to our normal visual perception. The
simple demonstration that an animal can respond to a moving target
after an ablation of a particular area is not sufficient to indicate
that area does not normally participate in movement perception and
strongly suggests that a large variety of stimulus dimensions and
neural areas participate in movement perception. The elimination
of one or another of these dimensions or areas would not necessarily
eliminate the overall perception of movement, as has been amply
demonstrated in form vision studies (Berkley and Sprague, 1979).
Despite these difficulties, there is some hope that in the not too
distant future more definitive conclusions will be possible with
regard to identifying the locale of the neural substrate of movement perception. In particular, the experiments of Zeki (1974a,
1974b) and others which seem to indicate that the nervous system
performs a second order abstraction of stimulus dimensions and then
processes that information in a unique neural locus appears to be
a valid idea. What is lacking, at the moment, are careful psychophysical studies of movement which each suspected area participates
in movement perception can be made.

The clinical studies of humans with striate cortex lesions,
have shown that these individuals can respond to moving targets
yet they deny, perceptually, that they see anything. Can we conclude then that the striate cortex is not necessary for the percep-

tion of movement? The clinical case reported by Heywood and Ratcliff (1975) in which a human with a unilateral colliculectomy reportedly responds to movement but reports that the scene appears to move in conjunction with normal head and eye movements, suggests that the superior colliculus is providing the long suspected corollary or suppression discharge for distinguishing self-induced movement from real movement.

The recent anatomical details that have been developed on the visual system also suggest rather broad representation of input from the retina to a variety of CNS structures. This broad input can distribute information about movement to numerous structures, and provides an anatomic rationale for findings of behavioral and electrophysiological studies which have suggested a model of movement perception which has many neural sites.

Further confirmation and details of the precise nature of the participation of these various neural areas await more detailed study both by investigators working on the psychophysics of movement perception and on the physiology and anatomy of the nervous system.

ACKNOWLEDGEMENTS

The technical assistance of D.S. Warmath and secretarial aid of Charlsine Mollica is gratefully acknowledged.
Supported by grants EY00953 and EY002259 from the National Eye Institute.

REFERENCES

Anderson, K.V. and Symes, D., The superior colliculus and higher visual functions in the monkey. Brain Res., 1969, 13, 37-52.

Baker, J., Gibson, A., Glickstein, M. and Stein, J., Visual cells in the pontine nuclei of the cat. J. Physiol., 1976, 255, 415-434.

Barlow, H.B. and Levick, W.R., The mechanism of directionally selective units in rabbit's retina. J. Physiol., 1965, 178, 477-504.

Baumgartner, G., Brown, J.L. and Schulz, A., Visual motion detection in the cat. Science, 1964, 146, 1070-1071.

Berkley, M.A., The role of the geniculo-striate system in vision. In F.A. King (ed.) Handbook of behavioral neurobiology, Vol. 1: Sensory Integration, R.G. Masterton, ed., New York: Academic Press, 1976, 63-120.

Berkley, M.A., Sherman, S.M., Warmath, D.S. and Tunkl, J.T., Visual capacities of adult cats which were reared with a lesion in the retina of one eye and the other occluded. Soc. Neurosci. Abstr., 1978a, 4, 467.

Berkley, M.A. and Sprague, J., Behavioral analysis of the role of
    geniculocortical system in form vision. In: Cool., E. and
    Smit, E. (eds.), Frontiers in Visual Science. New York:
    Springer-Verlag, 1977, 220-239.
Berkley, M.A. and Sprague, J.M., Striate cortex and visual acuity
    functions in the cat. J. Comp. Neurol., 1979, 187, 679-702.
Berkley, M.A., Warmath, D. and Tunkl, J.T., Movement discrimination
    capacities in the cat. J. Comp. Physiol. Psych., 1978b, 92,
    463-473.
Beverley, K.I. and Regan, D., Evidence for the existence of neural
    mechanisms selectively sensitive to the direction of movement
    in space. J. Physiol., 1973, 235, 17-29.
Blakemore, C. and Campbell, F., On the existence of neurones in the
    human visual system selectively sensitive to the orientation
    and size of retinal images. J. Physiol., 1969, 203, 237-260.
Bridgeman, B., Visual receptive fields sensitive to absolute and
    relative motion during tracking. Science, 1972, 178, 1106-1108.
Bridgeman, B., Receptive fields in single cells of monkey visual
    cortex during visual tracking. Intern. J. Neurosci., 1973a, 6,
    141-152.
Bridgeman, B., Background activity in single cells of monkey visual
    cortex during visual tracking. Intern. J. Neurosci., 1973b, 5,
    153-158.
Buchtel, H.A., Rubia, F.J. and Strata, P., Cerebellar unitary re-
    sponses to moving visual stimuli. Brain Res., 1973, 50,
    463-466.
Camisa, J., Blake, R. and Levinson, E., Visual movement perception
    in the cat is directionally selective. Exp. Brain Res., 1977,
    29, 429-432.
Campbell, F.W. and Robson, J.G., Application of fourier analysis to
    the visibility of gratings. J. Physiol., 1968, 197, 551-565.
Creutzfeld, O.D., Kuhnt, U. and Benevento, L.A., An intracellular
    analysis of visual cortical neurones to moving stimuli:
    Responses in a co-operative neuronal network. Exp. Brain Res.,
    1974, 21, 251-274.
Cynader, M., Berman, N. and Hein, A., Cats raised in a one-direc-
    tional world: Effects on receptive fields in visual cortex and
    superior colliculus. Exp. Brain Res., 1975, 22, 267-280.
Cynader, M. and Chernenko, G., Abolition of direction selectivity
    in the visual cortex of the cat. Science, 1976, 193, 504-505.
Cynader, M. and Regan, D., Neurones in cat parastriate cortex
    sensitive to the direction of motion in three-dimensional
    space. J. Physiol., 1978, 274, 549-569.
Daw, N.W. and Wyatt, H.J., Raising rabbits in a moving visual
    environment: An attempt to modify directional sensitivity in
    the retina. J. Physiol., 1974, 240, 309-330.
Daw, N.W. and Wyatt, H.J., Kittens reared in a unidirectional
    environment: Evidence for a critical period. J. Physiol.,
    1976, 257, 155-170.

Denny-Brown, D. and Chambers, R.A., Physiological aspects of visual perception. I. Functional aspects of visual cortex. Arch. Neurology, 1976, 33, 219-227.

Dews, P.B. and Wiesel, T.N., Consequences of monocular deprivation on visual behavior in kittens. J. Physiol., 1970, 206, 437-455.

Enroth-Cugell, C. and Robson, J.G., The contrast sensitivity of retinal ganglion cells of the cat. J. Physiol., 1966, 187, 517-552.

Gibson, A., Baker, J., Mower, G. and Glickstein, M., Cortico-pontine cells in area 18 of the cat. J. Neurophysiol., 1978, 41, 484-495.

Glickstein, M. and Gibson, A.R., Visual cells in the pons of the brain. Scientific American, 1976.

Glickstein, M., King, R.A. and Stein, J.F., The visual input to neurones in n. pontis. J. Physiol., 1971, 218, 79P.

Glickstein, M., Stein, J. and King, R.A., Visual input to the pontine nuclei. Science, 1972, 178, 1110-1111.

Goldberg, M.E. and Wurtz, R.H., Activity of superior colliculus in behaving monkey. I. Visual receptive fields of single neurons. J. Neurophysiol., 1972a, 35, 542-559.

Goldberg, M.E. and Wurtz, R.H., Activity of superior colliculus in behaving monkey. II. Effect of attention on neuronal responses. J. Neurophysiol., 1972b, 35, 560-574.

Graham, C. Perception of movement. In Graham, C. (ed.) Vision and Visual Perception., New York: Wiley, 1965, 574-588.

Grüsser, O.J., Grüsser-Cornellis, V., Neuronal mechanisms of visual movement perception and some psychophysical and behavioral correlates. In Jung, R. (ed.), Handbook of Sensory Physiology, VII/3, Part A. Springer Verlag, Berlin, 1973, 333-430.

Guillery, R.W., Patterns of fiber degeneration in the dorsal lateral geniculate nucleus of the cat following lesions in the visual cortex. J. Comp. Neurol., 1967, 130, 197-222.

Hamilton, C.R. and Lund, J.S., Visual discrimination of movement: Midbrain or forebrain? Science, 1970, 170, 1428-1430.

Hammond, P., Directional tuning of complex cells in area 17 of the feline visual cortex. J. Physiol., 1978, 285, 479-491.

Hammond, P. and MacKay, D.M., Response of cat visual cortical cells to kinetic contours and static noise. J. Physiol., 1975, 252, 43P.

Hammond, P. and MacKay, D.M., Differential responsiveness of simple and complex cells in cat striate cortex to visual texture. Exp. Brain Res., 1977, 30, 275-296.

Hammond, P. and MacKay, D.M., Modulation of simple cell activity in cat by moving textured backgrounds. J. Physiol., 1978, 284, 117P.

Held, R., IV. Dissociation of visual functions by deprivation and rearrangement. Psychologische Forschung, 1967-68, 31, 338-348.

Hess, R. and Wolters, W., Responses of single cells in cat's lateral geniculate nucleus and area 17 to the velocity of moving visual stimuli. Exp. Brain Res., 1979, 34, 273–286.

Heywood, S. and Ratcliff, G., Long-term oculomotor consequences of unilateral colliculectomy in man. In Lennestrand, G., and Bach-y-Rita, P., Basic Mechanisms of Ocular Motility and New Chemical Implications, Pergommon Press, 1975.

Höllander, H., Autoradiographic evidence for a projection from the striate cortex to the dorsal part of the lateral geniculate nucleus in the cat. Brain Res., 1972, 41, 464–466.

Hubel, D.H. and Wiesel, T.N., Receptive fields, binocular inter-action and functional architecture in the cat's visual cortex. J. Physiol., 1962, 160, 106–154.

Humphrey, N.K. and Weiskrantz, L., Vision in monkeys after removal of the striate cortex. Nature, 1967, 215, 595–597.

Johansson, G., Visual event perception. In Held, R., Leibowitz, H., and Teuber, H.-L. (eds.) Handbook of Sensory Physiology; Vol. VIII. Perception. Springer-Verlag; New York, 1978, 675–712.

Jones, K.R., Berkley, M.A., Spear, P. and Tong, J., Visual capaci-ties of monocularly deprived cats after reverse lid suture and enucleation of the non-deprived eye. Soc. Neurosci. Abstr., 1978, 4, 475.

Kawamura, S., Sprague, J.M. and Niimi, K., Corticofugal projections from the visual cortices to the thalamus, pretectum and superior colliculus in the cat. J. Comp. Neurol., 1974, 158, 339–362.

Kennedy, J.L., The nature and physiological basis of visual move-ment discrimination in animals. Psychol. Rev., 1936, 43, 494–521.

Kennedy, J.L., The effect of complete and partial bilateral extir-pation of the area striata on visual movement discrimination in the cat. Psychol. Bull., 1936, 33, 754.

Kennedy, J.L., The effects of complete and partial occipital lobectomy upon thresholds of visual real movement discrimina-tion in the cat. J. Genetic Psychol., 1939, 54, 119–149.

Kennedy, J.L. and Smith, K.U., Visual thresholds of real movement in the cat. J. Genetic Psychol., 1935, 46, 470–476.

Koerner, F. and Teuber, H.-L., Visual field defacts after missile injuries to the geniculostriate pathway in man. Exp. Brain Res., 1973, 18, 88–113.

Kozak, W., Rodieck, R.W. and Bishop, P.O., Responses of single units in lateral geniculate nucleus of cat to moving visual patterns. J. Neurophysiol., 1965, 28, 19–47.

Lange, H., de, Relationship between critical flicker-frequency and a set of low-frequency characteristics of the eye. J. Opt. Soc. Amer., 1954, 44, 380–389.

Lehmkuhle, S., Kratz, K.E., Mangel, S.C. and Sherman, S.M., Spatial and temporal sensitivity of X- and Y-cells in dorsal lateral geniculate nucleus of the cat. J. Neurophysiol., 1980, 43, 2, 520–541.

Lennie, P., Parallel visual pathways: A review. Vision Res., 1980, 20, 561-594.

Levick, W.R., Receptive fields and trigger features of ganglion cells in the visual streak of the rabbit's retina. J. Physiol., 1967, 188, 285.

Levick, W.R., Oyster, C.W. and Takahashi, E., Rabbit lateral geniculate nucleus: Sharpener of directional information. Science, 1969, 165, 712-714.

Maffei, L., Fiorentini, A. and Bisti, S., Neural correlate of perceptual adaptation to gratings. Science, 1973, 182, 1036-1038.

Mays, L.E. and Sparks, D.L., Dissociation of visual and saccade-related responses in superior colliculus neurons, J. Neuro-physiol., 1980, 43, 207-232.

Meyers, B. and McCleary, R.A., Interocular transfer of a pattern discrimination in pattern deprived cats. J. Comp. Physiol. Psychol., 1964, 57, 16-21.

Montero, V.M. and Brugge, J.F., Direction of movement as the significant stimulus parameter for some lateral geniculate cells in the rat. Vision Res., 1969, 9, 71-88.

Monty, R.A. and Senders, J.W. (eds.) Eye Movements and Psychological Processes., Lawrence Erlbaum Assoc., Publishers, 1976.

Movshon, J.A., Velocity preferences of simple and complex cells in the cat's striate cortex. J. Physiol., 1974, 242, 121-123.

Movshon, J.A., The velocity tuning of single units in cat striate cortex. J. Physiol., 1975, 249, 445-468.

Orban, G.A. and Callens, M., Influence of movement parameters on area 18 neurones in the cat. Exp. Brain Res., 1977, 30, 125-140.

Orban, G.A., Kennedy, H., Maes, H. and Amblard, B., Cats reared in stroboscopic illumination: Velocity characteristics of area 18 neurons. Arch. Ital. Biol., 1978, 11b, 413-419.

Pasternak, T. and Merigan, W.H., Abnormal visual resolution of cats reared in stroboscopic illumination. Nature, 1979, 280, 313-314.

Pasternak, T., Movshon, J.A. and Merigan, W.H., Motion mechanisms in strobe reared cats: Psychophysical and electrophysiological measures. Acta Psychologica, 1981, 48, 321-331. Special Issue on the Perception of Motion.

Peck, C.K., Crewther, S.G. and Hamilton, C.R., Partial interocular transfer of brightness and movement discrimination by split-brain cats. Brain Res., 1979, 163, 61-75.

Pettigrew, J.D., Nikara, T. and Bishop, P.O., Responses to moving slits by single units in cat striate cortex. Exp. Brain Res., 1968, 6, 373-390.

Pöppel, E., Held, R. and Frost, D., Residual visual function after brain wounds involving the central visual pathways in man. Nature, 1973, 243, 295-296.

Regan, D. and Beverley, K.I., Visual guided locomotion: Psycho-
    physical evidence for a neural mechanism sensitive to flow
    patterns. Science, 1979, 205, 311-313.
Regan, D. and Cynader, M., Neurons in area 18 of cat visual cortex
    selectively sensitive to changing size: Nonlinear interactions
    between responses to two edges. Vision Res., 1979, 19, 699-711.
Richards, W., Motion detection in man and other animals. Brain,
    Behav., and Evol., 1971, 4, 162-181.
Richards, W. and Smith, R.A., Midbrain as a site for the motion
    after-effect. Nature, 1969, 223, 533-534.
Riddoch, G., Dissociation of visual perceptions due to occipital
    injuries, with special reference to appreciation of movement.
    Brain, 1917, 40, 15-57.
Riesen, A.H. and Aarons, L., Visual movement and intensity discrimi-
    nation in cats after early deprivation of pattern vision.
    J. Comp. Physiol. Psychol., 1959, 52, 142-149.
Robinson, D.L. and Jarvis, C.D., Superior colliculus neurons studied
    during head and eye movements of the behaving monkey.
    J. Neurophysiol., 1974, 37, 533-540.
Robinson, D.L. and Wurtz, R.H., Use of an extraretinal signal by
    monkey superior colliculus neurons to distinguish real from
    self-induced stimulus movement. J. Neurophysiol., 1975, 39,
    852-870.
Rodieck, R.W., The Vertebrate Retina. Principles of Structure and
    Function. W.H. Freeman and Company, San Francisco, 1973.
Rodieck, R.W. and Stone, J., Response of cat retinal ganglion cells
    to moving visual patterns. J. Neurophysiol., 1965, 28, 819-832.
Rowe, M.H. and Stone, J., Naming of neurones: Classification and
    naming of cat retinal ganglion cells. Brain, Behav. Evol.,
    1977, 14, 185-216.
Schiller, P.H. and Koerner, F., Discharge characteristics of single
    units in superior colliculus of the alert rhesus monkey.
    J. Neurophysiol., 1971, 34, 920-936.
Schiller, P.H., Stryker, M., Cynader, M. and Berman, N., Response
    characteristics of single cells in the monkey superior colli-
    culus following ablation or cooling of visual cortex.
    J. Neurophysiol., 1974, 37, 181-194.
Schiller, P.H., True, S.D. and Conway, J.L., Effects of frontal eye
    field and superior colliculus ablations on eye movements.
    Science, 1979, 206, 590-592.
Sekuler, R., Pantle, A. and Levinson, E., Physiological basis of
    motion perception. In Held, R., Leibowitz, H.W., and Teuber,
    H.-L. (eds.) Handbook of Sensory Physiology, VIII, Springer
    Verlag, Berlin, 1978, 67-96.
Singer, W. and Bedworth, N., Inhibitory interaction between X and
    Y units in the cat lateral geniculate nucleus. Brain Res.,
    1973, 49, 291-307.
Singer, W. and Bedworth, N., Correlation between the effects of brain
    stem stimulation and saccadic eye movements on transmission in
    the cat lateral geniculate nucleus. Brain Res. 1978, 72, 185-202.

Smith, K.U., Experiments on the neural basis of movement vision. J. Exp. Psychol., 1941, 28, 199–216.

Sterling, P. and Wickelgren, B.G., Visual receptive fields in the superior colliculus of the cat. J. Neurophysiol., 1969, 32 1.

Sterling, P. and Wickelgren, B.G., Function of the projection from the visual cortex to the superior colliculus. Brain Behav. Evol., 1970, 2, 210–218.

Stone, J. and Dreher, B., Projection of X- and Y-cells of the cat's lateral geniculate nucleus to areas 17 and 18 of visual cortex. J. Neurophysiol., 1973, 36, 551–567.

Sutherland, N., Figural aftereffects and apparent size. Quart. J. Exp. Psych., 1961, 13, 222–228.

Tretter, F., Cynader, M. and Singer, W., Modification of direction selectivity of neurons in the visual cortex of kittens. Brain Res., 1975, 84, 143–149.

Vautin, R.G. and Berkley, M.A., Responses of single cells in cat visual cortex to prolonged stimulus movement: Neural correlates of visual aftereffects. J. Neurophysiol., 1977, 40, 1051–1065.

Vital-Durand, F. and Jeannerod, M., Role of visual experience in the development of optokinetic response in kittens. Exp. Brain Res., 1974, 20, 297–302.

Walls, G.L., The Vertebrate Eye. Cranbrook Institute Press, Bloomfield Hills, 1942.

Weiskrantz, L., Warrington, E.K., Sanders, M.D. and Marshall, J., Visual capacity in the hemianopic field following a restricted occipital ablation. Brain, 1974, 97, 709–728.

Wickelgren, B.A. and Sterling, P., Influence of visual cortex on receptive fields in the superior colliculus of the cat. J. Neurophysiol., 1969, 32, 16.

Wiesel, T.N. and Hubel, D.H., Single cell responses in striate cortex of kittens deprived of vision in one eye. J. Neurophysiol., 1963, 26, 1003–1017.

Wiesel, T. and Hubel, D.H., Comparison of the effects of unilateral and bilateral eye closure on cortical unity responses in kittens. J. Neurophysiol., 1965, 28, 1029–1040.

Wurtz, R.H., Comparison of effects of eye movements and stimulus movements on striate cortex neurons of the monkey. J. Neurophysiol., 1969, 32, 978–994.

Wyatt, H.J. and Daw, N.W., Directionally sensitive ganglion cells in the rabbit retina: Specificity for stimulus direction, size, and speed. J. Neurophysiol., 1975, 38, 613–626.

Zeki, S.M., Functional organization of a visual area in the posterior bank of the superior temporal sulcus of the rhesus monkey. J. Physiol., 1974a, 236, 549–573.

Zeki, S.M., Cells responding to changing image size and disparity in the cortex of the rhesus monkey. J. Physiol., 1974b, 242, 827–841.

IMPLICATIONS OF RECENT DEVELOPMENTS IN DYNAMIC SPATIAL ORIENTATION
AND VISUAL RESOLUTION FOR VEHICLE GUIDANCE[1]

H.W. Leibowitz[2], R.B. Post[3], Th. Brandt[4] and J. Dichgans[5]

INTRODUCTION

The objective of the present chapter is to examine a number of
problems of vehicle guidance within the context of recent developments in psychophysics, neurophysiology, anatomy and neurology.
The theme of this presentation is that these, like other applied
problems, exist primarily when the underlying physiological mechanisms are not well understood. With an increased understanding of
fundamentals we can not only better appreciate the nature of applied
problems but can also frequently identify methods towards their
solution. In turn, applied problems serve a valuable function by
directing attention to gaps in our basic understanding which leads
to suggestions for fruitful areas for research. At the same time as

[1] Sponsored by grants MH08061 from the National Institute of Mental
Health and EY03276 from the National Eye Institute, and by a senior
scientist award from the Alexander von Humboldt Foundation to
H.W. Leibowitz for study at the University of Freiburg, West Germany.
The authors are grateful to Jane Raymond and Eileen Leibowitz for
critically reading the manuscript.

[2] Department of Psychology, Moore Building, Pennsylvania State University, University Park, PA 16802.

[3] Department of Ophthalmology, University of California, Davis.

[4] Department of Neurology, Essen City Hospital, Essen, West Germany.

[5] Department of Neurology, University of Tübingen, Tübingen,
West Germany.

we learn more about fundamentals, new application possibilities
manifest themselves.

The plan of the present chapter is to review selectively three
areas in which recent research has increased our understanding of
fundamental mechanisms. These are 1) the two modes of processing
concept; 2) the multisensory nature of spatial orientation; and
3) the intermediate dark-focus of accommodation. Space does not
permit an exhaustive treatment of these topics. Rather, the
literature will be reviewed briefly in order to provide a basis for
describing the relevance of these developments to specific problems
in vehicle guidance and locomotion. *By juxtaposing basic developments and potential applications their close relationship should be
apparent, and the value of an approach which purposely de-emphasizes
differences between "basic" and "applied" science will, it is hoped,
be demonstrated.*

1. THE TWO MODES OF PROCESSING CONCEPT

The principal research stimulating interest in the two modes
of processing concept was Gerald Schneider's doctoral dissertation
at MIT (Schneider, 1967) which examined the effects of selective
ablation on visual capacities of the golden hamster. His research
indicated that cortical lesions interfered with the ability to make
visual pattern discriminations but did not affect spatial orientation. Conversely, lesions of the superior colliculus eliminated the
ability to actively orient in space, but if the animal was correctly oriented, there was no difficulty in making a pattern discrimination. These data were conceptualized within the framework of "two
visual systems" and this concept was later extended to include the
behavior of both human and nonhuman primates. In particular, Richard
Held has been active in interpreting sensory and sensory-motor phenomena in these terms (Held, 1968). Because the anatomy of the
human is so much more complex than the hamster, Held has suggested
that we refer to "two modes of processing" (Held, 1970). This is
not meant to imply that anatomy is not important, but rather that
at this stage it is preferable to concentrate on the conceptual
distinction suggested by Schneider's ablation studies.

The two modes of processing concept posits two independent and
dissociable modes of processing visual information. The "focal"
mode is concerned with object discrimination and identification, or
more generally, the question of "what?". It is subserved primarily
by the cortex and is typically well represented in consciousness.
Because focal functions involve higher spatial frequencies, they
are optimal in the central visual field and are systematically
related to both luminance and refractive error. The other mode of
processing, referred to as "ambient" vision, is concerned with
spatial orientation or, more generally, the question of "where?".
The properties of the ambient mode differ along many dimensions in
comparison with the focal mode. Although spatial orientation is

certainly possible, if not superior, in the central visual field (the main domain of the focal mode), it is also adequate with stimulation of the peripheral retina in spite of its coarse resolution properties. Hence low spatial frequencies are sufficient for processing information in the ambient mode and ambient functions are less sensitive to both refractive error and luminance than focal functions. For example, the radial localization of a single visual stimulus is unaffected by its energy content or by refractive error (Leibowitz et al., 1955; Post and Leibowitz, 1980). Circular vection, the illusory sensation of self-motion induced by moving the visual environment, is only slightly influenced over a wide range of stimulus parameters as compared with focal capacities (Leibowitz, Shupert-Rodemer and Dichgans, 1979; Post, 1982). With respect to consciousness, the ambient system is often poorly represented although by directing attention one can be aware of ambient activity.

A number of recent ablation studies as well as observations of brain damaged humans have suggested that it is possible for some orientation ability to be spared despite loss of focal vision. Weiskrantz has referred to this phenomenon as "blindsight", and there is an interesting and provocative literature on this topic (Pöppel et al., 1973; Perenin and Jeannerod, 1975, 1978; Weiskrantz et al., 1974). For the present purpose, the demonstration that it is possible to walk while reading demonstrates the dissociability of and some basic characteristics of focal and ambient functions. Even though attention is dominated by the reading material, orientation in space is carried out confidently and accurately on the basis of peripheral stimuli at an unconscious or subconscious level. If illumination is lowered or the retinal image blurred, the ability to read is degraded but orientation is relatively unimpaired. A recent summary of the two modes of processing concept points out how a number of interesting problems in both basic and applied science can be fruitfully interpreted in terms of these emerging concepts (Leibowitz and Post, 1982).

Unlike vestibular stimuli, which normally lead to the sensation of body motion, visual motion stimuli allow for two perceptual interpretations, either object- or self-motion. A subject watching moving stimuli may perceive himself as being stationary in space (egocentric motion perception) or will experience the actually moving surroundings as being stable and himself as being moved. The latter illusion has been known for a long time but has not been thoroughly studied under experimental conditions until recently (for a review see Dichgans and Brandt, 1978). Visual perception of self-motion is dependent both on the density of moving contrasts within the visual field and the total area of the stimulus. Additionally, with simultaneous presentation of foreground and background optokinetic stimuli, dynamic visual spatial orientation relies mainly on the information from the background (Brandt, et al., 1975). An analogous situation exists while riding in a car at constant velocity where self-motion sensation is produced by a relative backward motion of the surroundings rather than the stable image of the car.

The driver looking in the rear view mirror is able to detect and to pursue single objects with respect to himself and in relation to the environment without affecting the sensation of self-motion.

A critical problem in vehicle guidance which can be understood in terms of this theoretical approach is the high frequency of night driving accidents (Leibowitz and Owens, 1977). Automobile accidents, of course, have multiple causes. The role of illumination is demonstrated, however, by studies which indicate that when other factors are held constant, accidents increase dramatically when illumination is lowered, particularly those involving collisions with pedestrians (Gramberg-Danielsen, 1967). It is well known that under twilight and nighttime conditions many visual capacities such as spatial resolution, stereoscopic depth perception, contrast discrimination, and reaction time are degraded. This is reflected in analyses of nighttime accidents in which drivers frequently report that they did not see a pedestrian or other obstacle in time to stop. In some cases, the sound of impact was heard before the driver was aware of the pedestrian (Allen, 1970). It is not necessary to read the visual literature in order to appreciate the fact that we simply do not see as well at night as during the day. What is curious, however, is that drivers typically do not reduce their speed at night, even though they are most probably aware through personal experience, or even through knowledge of the literature, that their vision has been degraded.

A possible explanation for this paradox may be derived from the two modes of processing concept (Leibowitz and Owens, 1977). Driving an automobile, in common with locomotion and vehicle guidance in general, involves two parallel tasks. Spatial orientation is accomplished by steering the vehicle, which requires continuous evaluation of the location of the vehicle relative to the road. In terms of the two modes concept, steering is concerned with "where" and is an ambient function. Driving also involves focal vision, the role of which is to monitor the roadway ahead for pedestrians, other vehicles and obstacles, to read traffic signs and monitor signals, and to judge the distance and speed of other vehicles. During daylight both the ambient and focal modes are operating at their maximal capacities. However, under twilight and nighttime conditions there is a selective degradation of the two modes. Focal visual functions are degraded, e.g., contrast sensitivity is diminished and spatial and stereoscopic acuity are reduced. (For many individuals the ability to appreciate detail is further degraded by a condition known as "night myopia" which will be discussed later in this chapter.) On the other hand, the efficiency of ambient visual functions is not reduced by lowered illumination. As long as minimal visual stimulation is available, it is possible to steer the vehicle adequately. As a result of the continuous feedback regarding this ability to steer the vehicle, the driver is not aware that there is a problem with degradation of focal vision, and therefore typically maintains the same velocities at night as during the day. In terms of the performance information available to the driver, it is the ambient mode which dominates. Since the demands on focal vision are intermittent, its degradation is only rarely reflected in the operator's performance. The result of this selective degradation of visual function is that the driver's self-confidence is inappropriate for the degraded focal visual function. Adding to the tendency to disregard focal degradation is the fact that critical incidences involving focal vision while driving are rare, their perceived risk is correspondingly low, and they tend to be ignored (Slovic et al., 1978). The net result of these factors is that drivers are typically unaware of focal degradation, rely on the efficiency of the ambient processes, and drive too fast to recognize and respond to infrequent hazards.

As is often the case, understanding the basic cause of a problem suggests methods for amelioration. In the case of the high nighttime accident rate a number of possibilities are apparent. Obviously, illuminating highways would be expected to be effective and this is supported by empirical observations. However, economic considerations limit this possibility. Other alternatives are to post different maximum velocities for night and daytime driving conditions. Before the introduction of the uniform national 55 mph speed limit in the United States, only a few states followed this practice, usually on major highways. To our knowledge, different speed limits have not been posted in areas where degradation of focal vision would be expected to play a major role in accidents with pedestrians. Another possible measure is to educate drivers regarding the potential dangers associated with selective degradation of vision at night. This procedure would be expected to be particularly effective for younger drivers whose habits have not been established. The implications of selective degradation should also be communicated to pedestrians and cyclists who should be encouraged to take special measures to increase their visibility at night in order to compensate for the loss of focal vision among drivers.

Another way to view the night driving accident problems is in terms of self-confidence of the driver. It is not the hazards and dangers of which we are aware which cause the most difficulty but rather those of which we are not aware. Unjustified self-confidence is often cited as the mechanism accounting for the increase in accidents following alcohol consumption. Alcohol is known to increase self-confidence even at concentrations which do not affect motor performance. Considering the fact that the selective degradation of vision at night also produces an unjustified sense of self-confidence, the gruesome data on nighttime accidents involving drinking are somewhat more understandable.

It is relevant to point out that the initial event in this discourse concerned Schneider's ablation studies in the hamster. Applications to night driving accidents played no role in motivating his research. That Schneider's fundamental data have led to increased understanding and, it is hoped, reduction of nighttime driving accidents illustrates the intimate relationship between basic re-

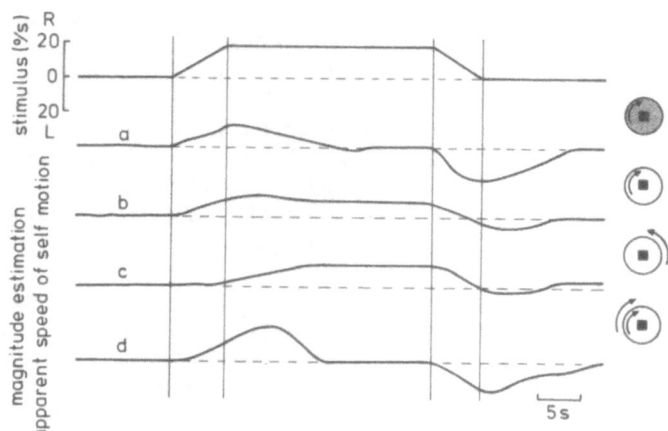

Fig. 1. Original recording of continuous tracking of perceived
        self-motion velocity and direction during chair and/or
        surround motion (trapezoid velocity profile, top trace).
        During chair rotation in dark (A) velocity profile roughly
        follows mechanical characteristics of cupula-endolymph
        system, resulting in lack of constant velocity discrimina-
        tion and consequent misinterpretation of deceleration. With
        a visible surround providing adequate optokinetic informa-
        tion these deficiencies are largely compensated (B). Isolated
        visual effect is demonstrated in (C) where (with consider-
        able latency) apparent self-rotation is elicited in a sta-
        tionary observer through exclusive surround motion in oppo-
        site direction. If the visual surround moves with the ob-
        server, as in vehicles, (D) self-motion perception is again
        erroneous since, as in A, it relies exclusively on vestibu-
        lar inputs. After Dichgans and Brandt, 1978.

search and its societal applications.

2. THE MULTISENSORY NATURE OF SPATIAL ORIENTATION

     Spatial orientation is mediated by the coordination of motor
activity with multiple sensory inputs, primarily from the visual,
vestibular and somatosensory systems (Dichgans and Brandt, 1978).
Under everyday conditions, the maintenance of posture and orienta-

tion in space proceeds smoothly and accurately with minimal demands
on consciousness. However, under special conditions, such as are
encountered in vehicle guidance, the normal interaction among these
sensory inputs may be altered, thereby producing a "mismatch" in
comparison with the previous experience of the individual. This
intersensory mismatch can produce disorientation, motion sickness,
or both.

In order to appreciate the nature of this mismatch, it is helpful to analyze the isolated contributions of individual senses. For
the case of rotation, the sensations which result from visual or
vestibular stimulation either in isolation or combined are presented
in Figure 1. The sensations of body motion resulting from rotations
of a subject about the vertical axis in the dark are described in
panel A. Estimates of speed and direction of movement sensation in
the absence of vision are veridical only during the acceleration .
phase. During rotation at a constant velocity the sensation of self-
motion gradually diminishes. This would be expected from the mechanical properties of the cupola-endolymph system as would the report
of subjective motion in the direction opposite from actual rotation
during physical deceleration. After physical motion ceases, this
illusory self-motion persists temporarily. Note that in this case
the only spatial orientation information corresponding to rotation
is from the vestibular system, and sensations are veridical only
during the initial acceleration stage, and, after a delay, when the
subject is again at rest.

Illusory sensation results when the experimental situation
prevents the normal interaction between the visual and vestibular
senses. Neither vision alone nor the vestibular system alone can
provide accurate information. However, if normal interaction is
provided by rotating the subject in the light, the combined effect
of the visual and vestibular inputs is a relatively veridical report
of the direction and magnitude of self-motion (panel B).

The isolated contribution of vision can be demonstrated by rotating a surround, consisting of vertical stripes on the inside of
a drum, around the subject, so that the resulting retinal image
motion corresponds to that produced when the subject is rotated
(panel C). In this case, the initial subjective sensation is also
veridical, i.e., the subject reports surround motion in the correct
direction. However, after a latency of 3 to 5 seconds, illusory
self-motion is reported in the direction opposite from surround
motion. At first both surround motion and self-motion are experienced but the surround motion gradually diminishes and is replaced,
after 5 to 10 seconds, with exclusive self-motion. During this
".saturated" phase, the surround appears to be stationary and the
subject reports a compelling feeling of self-motion in the direction
opposite to the surround. During surround deceleration, apparent
self-motion or circular vection gradually diminishes but when the
surround is again at rest there is a tendency toward apparent self-
motion in the opposite direction, this effect being greater in the
dark. Thus, large field visual stimulation analogous to that pro-

duced when the body is moving in space produces, in the absence
of other sensory inputs, veridical spatial orientation only during
the initial few seconds and, after a latency, when the surround is
again at rest.

It should be noted that not only is combined stimulation from
the vestibular and visual senses necessary for accurate orientation,
but also that the sensory information must correspond with previous
experience. If the chair and surround motion are mechanically
coupled so that the surround is visible but moves with the subject
a mismatch occurs and, as was the case for isolated visual and
isolated vestibular stimulation, spatial disorientation is experienced after the initial acceleratory phase (panel D).

Functional relations highly similar to those for subjective
reports of self-motion have been obtained in recordings from the
vestibular nucleus of the goldfish (Dichgans, et al., 1973) and the
monkey (Waespe and Henn, 1976) under similar stimulus conditions.
These studies indicate that neural analogues for visual-vestibular
integration may be found in the brain stem.

Sensory interactions in spatial orientation and the consequences
of isolation of sensory inputs have long been recognized as a source
of orientation illusions in aircraft. It is well known that pilots
must disregard vestibular information when the ground is not visible
and rely exclusively on information from the flight instruments.
However, in spite of this emphasis, disorientation may occur even
among experienced pilots in commercial aviation (Benson, 1978). For
example, pilots have reported that perceived orientation (e.g.,
whether the wings are level) may be different from the information
indicated by the flight instruments. In some cases a conflict between sensations and the information from the instruments interferes
with the ability to control the aircraft.

An interpretation of the nature of this disorientation follows
from the basic mechanisms underlying normal spatial orientation
(Brandt & Daroff, 1980). In everyday life acceleration of the head
produces both forces which activate the vestibular system and motion
of the retinal images of surrounding objects. However, acceleration
in an aircraft, when outside detail is not visible, carries the
visual surroundings along with the head. As a consequence, forces on
the vestibular system are no longer matched by retinal image motions
of the surroundings. This represents a mismatch or conflict in comparison with previous experience which the pilot must "override" in
order to maintain correct orientation. (This is analogous to the
stimulus situation presented in Fig. 1, panel D.) This overriding
process requires the pilot to interpret the roll indicator instrument
and to ignore conflicting information from the vestibular and visual
senses. In this situation, orientation information is provided by a
small, usually 5° in diameter, instrument. This contrasts with the
normal situation in which visual orientation information is provided
by stimulation of large areas of the visual field and does not require an intermediate interpretation stage. In effect, a cognitive
skill learned later in life is in conflict with an automatic response

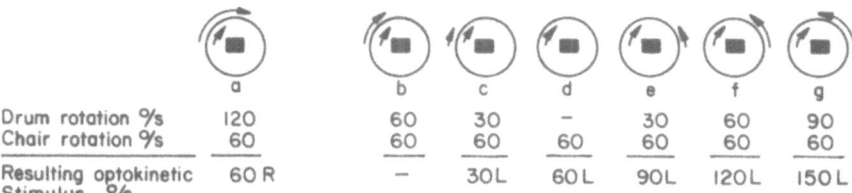

	a		b	c	d	e	f	g
Drum rotation °/s	120		60	30	–	30	60	90
Chair rotation °/s	60		60	60	60	60	60	60
Resulting optokinetic Stimulus °/s	60 R		–	30L	60 L	90L	120L	150 L

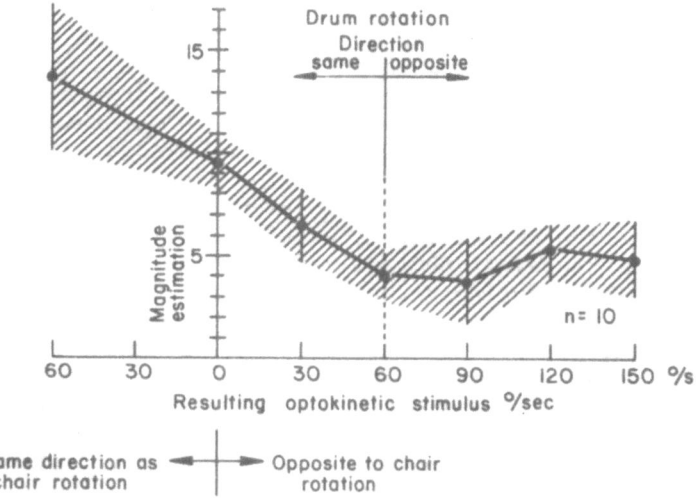

Fig. 2. Magnitude estimation of Coriolis effects, induced by head movements during constant velocity (60 deg/sec) chair rotation while direction and velocity of optokinetic stimulation provided by additional surround motion are varied (means and standard deviation). Stimuli are listed in upper part of figure. Weakest effects occur if motion of the surround is in physiological agreement with real body rotation (d): Optokinetic inhibition of "vestibular" Coriolis effects is similar for increase of velocity beyond 90 deg/ sec (e-g). Inhibition is reduced if velocity of relative backward motion lags actual chair rotation (c). Coriolis effects, however, are markedly enhanced if direction of circular vection is opposite to actual chair rotation (a). Direction-specific interaction of two stimuli is roughly linear within a range of ± 60 degrees of relative velocity (a-d). After Dichgans and Brandt, 1973.

which has predominated throughout the past history of the observer
(Leibowitz and Dichgans, 1980). Although in the majority of cases
pilots are able to concentrate on the information from the flight
instruments, there are incidences in which conflicts cannot be resolved and disorientation results.

From this analysis, it follows that any procedure which reduces
the mismatch should improve spatial orientation. Since the vestibular
inputs are essentially unalterable, the only practical possibility
involves presentation of visual information which is more similar
to that which normally subserves spatial orientation. Specifically,
visual information for orientation is typically available over a
large area of the visual field. Similarly, if information regarding
the roll attitude of the plane also stimulates a larger extent of
the visual field than the small area represented by a flight instrument, orientation should be facilitated. Malcolm et al. (1975) have
proposed such a device in which an extended artificial horizon is
projected on the windscreen. Preliminary reports indicate that this
wide artificial horizon is effective in maintaining spatial orientation (Money et al., 1976). These results are consistent with the
point of view that a major cause of disorientation is an intersensory mismatch as compared with previous experience.

Intersensory mismatch, particularly when severe, can also lead to
motion sickness. If the semicircular canals are moved out of their
normal planes during rotation about the vertical axis, a cross-cou-
pling or Coriolis effect results. The corresponding sensation is of
extreme disorientation and frequently produces motion sickness. The
effect of systematic variation of the correspondence between information signaling the direction of rotation from the vestibular and the
visual systems, while making head movements during rotation, is illustrated in Fig. 2. When the visual surround was stationary so that information regarding direction of rotation from the vestibular and
visual systems was in harmony, minimal discomfort, as evaluated by
magnitude estimation, was reported (d). If the drum was rotated in
the opposite direction from the subject so that the information from
the visual and vestibular senses still agreed with respect to direction of rotation, rated discomfort remained the same even when drum
velocity was increased (e-g). If, however, the drum rotated in the
same direction as the subject and with the same speed so that information from the two senses was in conflict, (i.e., vision indicated
that the subject was at rest while the vestibular sense signaled
rotation) estimation of discomfort doubled (b). The worst case
occurred when visual and vestibular information indicated rotation
in opposite directions, as illustrated in panel a. In this conflict
situation, several subjects became motion sick and vomited after
only one head movement.

An analogous example of intersensory mismatch within the context of vehicle guidance occurs in automobiles. For the driver who
has a relatively unobstructed view of the stable external world,
acceleration of the vehicle results in opposite movement of the
visual field which is consistent with previous experience (Fig. 3).

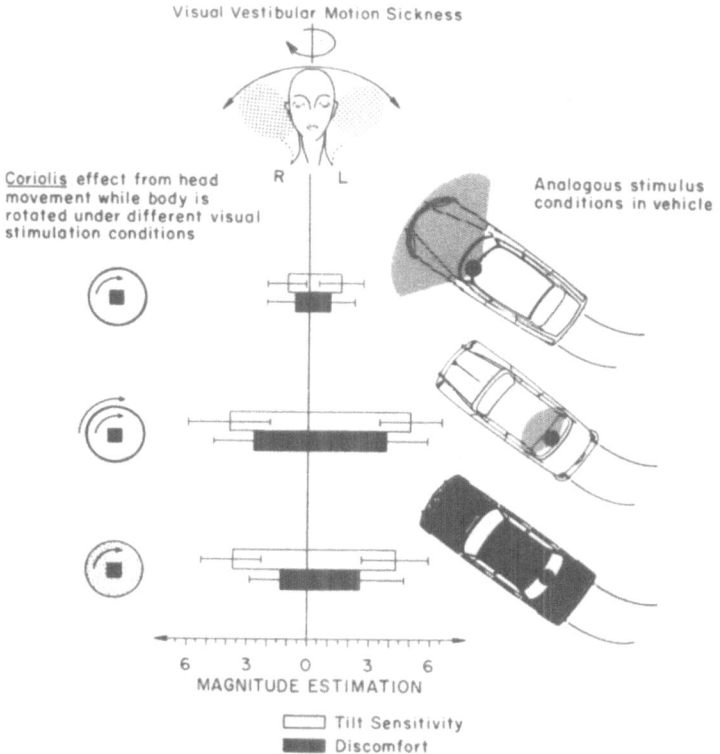

Fig. 3. Magnitude estimation of tilt sensations and discomfort resulting from sideward movement of the head (Coriolis stimulation) during simultaneous chair and visual surround
rotation. The symbols on the left, from top to bottom, represent: chair rotation in the light, coupled chair and
surround motion, and visual surround rotation only. The
right column represents comparable stimulus situations in
a moving vehicle. Motion sickness is minimal in the front
seat where body accelerations and simultaneous visual
stimulation are corresponding. Discomfort is maximal when
the vestibular accelerations are contradicted by visual
information indicating no movement as occurs in the back
seat with predominantly stationary contours in the visual
field. After Brandt, 1976.

However, for a passenger in the back seat of the vehicle, a relatively greater portion of the visual field is stimulated by the
interior of the vehicle so that during acceleration visual stimuli
fail to move in the appropriate (opposite) direction, which results
in a mismatch. This accounts for the fact that motion sickness is
typically more prevalent in the rear than the front seat of an automobile (Brandt, 1976). If one attempts to read in a vehicle, an even
larger proportion of the visual field produces a mismatch signal
and this further increases the tendency toward motion sickness.

Susceptibility to motion sickness has interesting developmental
aspects. Children below two years of age are highly resistant to
motion sickness but thereafter up to about twelve years of age
reveal a greater susceptibility than adults (Money, 1970). This has
been attributed to a preference for riding in vehicles in the supine
position (Chinn and Smith, 1955) but can be explained more convincingly by the finding that very young children apparently do not
suffer from visual-vestibular mismatches (Brandt, et al., 1976). The
visual contribution to the multiloop control of dynamic spatial
orientation emerges rather late in development and is sequentially
calibrated after acquisition of upright stance and gait.

The differential effects of visual parameters on motion sickness
in cars were tested under real road conditions for linear decelera-
tions (0.1 - 1.2 g) by exposing subjects to repetitive braking
maneuvers (Probst, et al., 1981). The severity of motion sickness
was found to depend on the visual stimulus conditions with significant differences among the following conditions: 1) with adequate
visual surround visible moderate nausea was experienced, 2) with
eyes closed and somatosensory-vestibular excitation only, medium
nausea was reported, and 3) with conflicting visual input so that
acceleration was in disagreement with the visual information of no
movement, strong nausea was experienced. In general, providing ample
full field vision of the relatively stationary surround is the best
strategy to alleviate car sickness (Brandt, 1981).

The contribution of intersensory mismatch to disorientation and
motion sickness is receiving increasingly more attention (Money,
1970; Reason and Brandt, 1975; Benson, 1978). The basic mechanisms
outlined here are consistent with many familiar observations. The
tendency toward motion sickness on a ship is reduced when stable
contours such as the horizon are visible. The contribution of past
experience in determining whether a particular pattern of sensory
stimulation will produce a mismatch is illustrated by the adaptation process. After an individual is repeatedly exposed to a new
pattern of intersensory matching, such as on a ship, the severity
of disorientation and motion sickness decrease. If the previous
pattern of interaction is reintroduced, a mismatch will result.
This is illustrated by the phenomenon of mal de débarquement, which
is often observed when leaving a ship after a long voyage. This
same phenomenon has been reported following space flight.

Under the microgravity of space flight, active head movements
elicit motion sickness. One explanation is that the symptoms are

generated by an intravestibular mismatch between the otoliths and semicircular canals (Benson, 1977). The afferent consequences of self-generated body movements in space are different from the signals expected from prior calibrations on earth. The semicircular canals and otoliths correctly transduce angular and linear accelerations, respectively, in space as on earth, but the otoliths fail to signal orientation of the head in the absence of gravity. A second possible mechanism for space sickness is related to the normal slight difference in absolute weights of the two otoliths (von Baumgarten and Thümler, 1970). This difference is compensated for by central mechanisms in the gravitational environment on earth. In the microgravity of space, a recalibration is required, and astronauts are symptomatic until it is accomplished.

The modifiability of the spatial orientation mechanisms is dramatically illustrated by studies of perceptual-motor adaptation to inversion of the retinal image (Kaufman, 1974). More recently, exposure to rearranged visual and vestibular inputs has been shown to result in alteration of the magnitude of and direction of the vestibulo-ocular reflex with even reversal of direction possible (Gonshor and Melville-Jones, 1976).

This analysis of spatial orientation in terms of the interaction among subsystems has resulted in new insights into disorientation and motion sickness. In this context, many problems of vehicle guidance are interpreted as resulting from exposure to patterns of sensory information which differ from those encountered in the normal environment to which the individual has become adapted during years of previous experience. New approaches to vehicle guidance problems follow from an increased understanding of the mechanisms which have evolved to subserve spatial orientation and locomotion.

The basic research which provided the data and concepts leading to our present understanding of spatial orientation derived from many sources. However, the majority of these efforts originated in laboratories associated with medical clinics in which the tradition of a close proximity between basic understanding and applications has been maintained. We do not consider it to be accidental that advances in medicine frequently originate from individuals who are both committed to basic research and responsible for the diagnosis and treatment of patients. Disabling vertigo·in patients is subserved by the same neurological systems as spatial disorientation in aircraft. The disturbances of normal function seen in clinics, unfortunate as they are, are natural "experiments" which test our understanding of normal function and frequently provide valuable guidelines for analysis and study.

3. ANOMALOUS MYOPIAS AND THE INTERMEDIATE DARK-FOCUS OF ACCOMMODATION

A long-standing problem in visual science with implications for vehicle guidance has been that many individuals apparently accommodate to near distances even when the object of interest is far away.

Such inappropriate accommodation is maladaptive and is referred to
as "anomalous myopia". The first report of an anomalous myopia was
made in 1789 by Lord Maskelyne, Director of the Royal Greenwich
Observatory, who noticed that he became nearsighted at night
(Levene, 1965). This condition, referred to as "night myopia", has
been a persistent problem for almost two centuries because it degrades the sharpness of the retinal image and interferes with the
ability to appreciate distant detail under twilight and nighttime
observation conditions. An analogous phenomenon occurs in daylight
if visual detail is absent. When searching the sky or during a fog
or snowstorm, the focus of many observers with normal distance
vision corresponds to an intermediate distance, thereby reducing
the probability of detecting small distant objects. Considerable
attention has been devoted to this phenomenon, referred to as
"space myopia" or "empty field myopia", in military aviation
(Whiteside, 1952).

About a decade ago, the availability of low-cost lasers made
feasible the construction of an optometer which accurately evaluates
accommodation without affecting its magnitude (Knoll, 1966; Baldwin
and Stover, 1968; Hennessy and Leibowitz, 1970). This instrument
can also be used in situations in which the anomalous myopias are
typically exhibited. Utilizing this new methodology, the validity
of the intermediate resting position or dark-focus was empirically
confirmed. Figure 4 represents the frequency distribution of the
accommodation distance of 220 college students in total darkness
(Leibowitz and Owens, 1975, 1978). All observers wore their normal
corrections during the observations. The average resting position
of this group is at an intermediate value of 1.52 diopters, corresponding to a distance of 65 cm (26 in). Only a few of the subjects
demonstrated the classical infinity resting position. The most
striking feature of these data, however, is the large intersubject

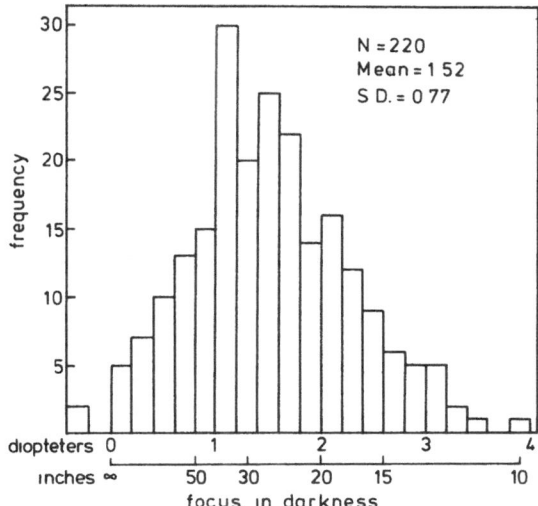

Fig. 4. Frequency distribution of the magnitude of the focus in
total darkness, as measured with the laser optometer, for
220 college-age observers. After Leibowitz and Owens, 1978.

variability. In contrast to the subjects who demonstrate an infinity
resting position, the dark-focus of others was as close as four
diopters or 25 cm (10 inches).

Demonstration of the validity of the intermediate as opposed
to the infinity dark-focus, together with the revelation of the
marked intersubject variability, has provided the key to both predicting and correcting the anomalous myopias. If one assumes, as
has been the case historically, that the resting position is at
infinity, then any accommodation to a distance nearer than the
object of interest must represent an active maladaptive process.
Alternatively, if the resting or tonus position of the accommodative
system corresponds to an intermediate position, then any interference with the efficiency of the accommodative feedback loop would
result in a <u>passive</u> return of accommodation toward the dark-focus.
Furthermore, given the large intersubject variability of the dark-
focus, anomalous myopias would also be expected to demonstrate
comparable individual differences. Those subjects with an infinity
dark-focus should not show any anomalous myopia, while those with
near dark-focus values would tend to exhibit anomalous myopia in
an amount related to their individually characteristic dark-focus
values. This line of reasoning has interesting consequences: 1) It

eliminates the paradox of an active maladaptive response for a
system which is normally highly efficient, 2) it accounts for the
failure of previous theoretical explanations to successfully handle
the problem of intersubject variability, and 3) it provides a
hypothesis which can easily be tested: the magnitude of anomalous
myopia should be predictable from the individual subject's dark-
focus and should be correlated for all viewing conditions in which
accommodation is degraded even though the stimulus conditions for
specific anomalous myopias differ. In the case of night myopia,
lowering illumination would be expected to interfere with accommodation because it restricts the ability of the neural structures
to process contrast in the retinal image. In the case of space
myopia, even though illumination is high, there are no focusable
contours available to provide the error signal necessary for the
accommodation reflex. In both cases, the anomalous myopia should
be proportional to the dark-focus.

     In a test of this hypothesis, the magnitude of anomalous myopia
was determined while: 1) viewing a distant target under twilight
illumination (night myopia), 2) observing a bright field without
contours (space myopia), and 3) while viewing a high contrast
target under bright illumination in a microscope (Leibowitz and
Owens, 1975). This latter condition, designed to produce "instrument myopia", has minor performance implications but is of interest
here because it represents still another example of anomalous
accommodation. In this case, the small exit pupil of the microscope
eliminates the need for accommodation. The magnitudes of anomalous
myopia obtained under these three viewing conditions are presented
in Figure 5 as a function of the dark-focus of the individual subjects. On these plots, if accommodation corresponds to the individual subject's dark-focus, the data should fall along the slanted
lines. It is apparent that the anomalous myopias can be largely predicted from the dark-focus values, and that the magnitude of the
anomalous myopia for individual subjects is the same for all three
viewing conditions. The agreement is better for night and empty
field observation conditions than for instrument viewing presumably
because the eyepiece of the microscope influences convergence which
in turn induces accommodation due to the normal coupling between
accommodation and convergence.

     With the ability to predict the magnitude of anomalous myopia
and to understand the underlying mechanism, the means for amelioration follow logically. In the case of night myopia, a number of
previous studies indicated that viewing through a negative lens
sometimes resulted in improvement. However, some subjects were not
helped at all while still others saw more poorly. In view of the
intersubject variability in the dark-focus, these results would now
be expected. Based on this variability, it was hypothesized that
any correction should be related to the individual subject's dark-
focus value. This assumption has been successfully tested in both
field and laboratory studies (Owens and Leibowitz, 1976) which
demonstrated the optimal visual correction for driving at night or

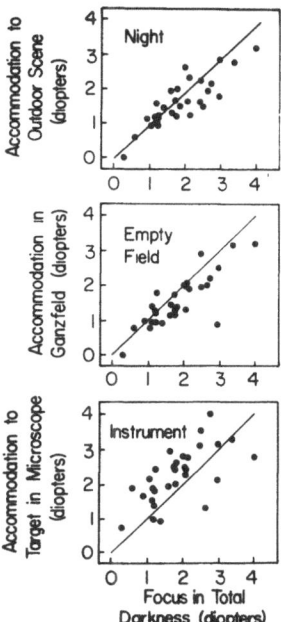

Fig. 5. Magnitudes of night, empty field, and instrument myopia as
a function of the focus in the dark. Each point represents
a datum for an individual subject who observed in all three
situations. After Leibowitz and Owens, 1975.

under simulated night driving conditions to be approximately one-
half of the individual's dark-focus value. In laboratory and field
studies based on this rule, all of the subjects tested so far with
a dark-focus nearer than infinity have demonstrated improvement in
their ability to appreciate detail under twilight and night time
observation conditions. Similar results have been reported in
laboratory studies of space myopia. In this case the detection of
a small target in a bright unstructured field is significantly improved by a correction which corresponds to the subject's dark-
focus value (Post et al., 1979; Luria, 1980). Within an aviation
context, a significant improvement in the sighting range for small
objects in the atmosphere would be expected with this correction.
The difference between the approximately half dark-focus correction
for night myopia and the full dark-focus correction for space
myopia follows naturally from the specific basis of accommodation
degradation. Under nighttime driving conditions the stimulus to
accommodation is reduced but not completely eliminated, but with

space myopia there is no stimulus to accommodation whatsoever, and accommodation returns completely to the dark-focus.

The empirical demonstration of the validity of the intermediate dark-focus and particularly the large intersubject variability has provided a simple yet effective basis for understanding and ameliorating the anomalous myopias. The general rule is that for visually demanding conditions, including but not limited to night driving and aviation, improvement in the quality of the retinal image will result from a spherical correction which shifts the distance at which no accommodative effort is required to the distance of the object of interest. This correction eliminates the need for active accommodation and the potentially deleterious effect of focusing errors resulting from interference with the accommodative feedback loop.

The history of the anomalous myopias illustrates the indispensible role of the understanding of fundamental mechanisms in dealing with practical problems. In this case, the obstacle to understanding the basis of anomalous accommodation was the incorrect assumption that relaxed accommodation typically corresponds to optical infinity. In spite of many discordant observations, not limited to the anomalous myopias, this view predominated in the literature for more than a century and led to unsuccessful attempts to understand these phenomena in terms of inappropriate mechanisms. Within the context of an intermediate resting position and its large intersubject variability, not only are the anomalous myopias understandable and correctable, but other visual problems both theoretical and applied, which previously appeared paradoxical, are now readily explained (Johnson, 1976; Owens, 1979).

4. SOME IMPLICATIONS FOR SELECTION, LICENSING, AND TRAINING

Tests for selection and licensing are formulated to match the capabilities of the vehicle operator with the capacities required to safely and effectively guide the vehicle. As our understanding of the nature of vehicle guidance tasks increases, we are in a better position to specify the relevant perceptual-motor skills, to evaluate the effects of factors such as aging, pathology, and drugs, and to recommend training procedures. A number of recent developments, particularly studies which demonstrate the modifiability of orientation systems, have important implications for these problems.

4.1. Night and space myopia

Night myopia can add to the degradation of focal vision which normally accompanies lowered illumination. The recent findings regarding the dark-focus of accommodation imply that vehicle driver licensing evaluations should include a test for night myopia since wearing a special prescription would be expected to substantially improve nighttime visual performance. Data on a university aged population in the United States (17-21 years) indicate that, based on an assumed nighttime refractive error equal to one-half the dark-focus, 76 percent of this group would be at least 0.5 diopter near-sighted at night, 26 percent would have a night refractive error of at least one diopter and four percent more than 1.5 diopters. The significance of these levels of optical blur can be appreciated when one considers that refractive errors of 0.25 to 0.5 diopter are typically corrected clinically and a noncorrectable spherical error of 2.5-3.0 diopters corresponds to legal blindness in the United States.

It should be pointed out that our current understanding of this problem does not permit an immediate translation into driver licensing procedures. Problems of individual vision and oculomotor characteristics, the effects of driving conditions and habits, and the role of experience and age remain to be analyzed clinically. However, at present we have strong evidence that a correction corresponding to approximately one-half the dark-focus is a good first approximation, as it has been shown to increase image clarity at night in both laboratory and field studies (Owens and Leibowitz, 1976). Such a correction should therefore contribute to the efficiency of foveal detail vision under reduced illumination conditions*.

A characteristic of the best pilots is superb distance vision. Any advantage in the ability to detect another aircraft at a greater range has obvious tactical importance. Since refractive errors resulting from space myopia impair the detection of stimuli subtending small visual angles, it follows that the focus of pilots while searching the sky for other aircraft should be at or very close to optical infinity. Any tendency away from an infinity focus would be expected to degrade performance in this situation. Three factors which would tend to result in a focus nearer than optical infinity are:

1. The natural population tendency toward an intermediate dark-focus.

2. The tendency of contours surrounding the windscreen to induce accommodation.

3. The effect of previous fixation on near stimuli, such as flight instruments, on the subsequent ability to focus for infinity.

These effects can be readily quantified and avoided either by selecting pilots with far dark-focus values or by an optical correction based on the individual's dark-focus.

---

* Refractive errors resulting from night myopia are not likely to affect visually induced self-motion perception (vection) since this involves large luminance contours the resolution of which would not be influenced by refractive error (Leibowitz et al., 1977). Such refractive errors would, however, be expected to affect object motion perception.

4.2. Individual differences in the effectiveness of spatial orienta-
     tion stimuli

Large intersubject variability is typically observed in experimental studies of spatial orientation. If it results from differences in the gain of the sensory systems subserving orientation, or
their interaction, it would have important implications. This latter
possibility is suggested by marked individual differences in susceptibility to motion sickness. The role of past experience in
"calibrating" the interaction among the sensory subsystems is also
relevant in this context.

At this stage, we are certain only of the fact that marked
individual differences exist. It remains to be determined whether
such data can be useful in predicting spatial disorientation and
motion sickness both of which have serious implications in aviation
and space flight.

4.3. Contrast sensitivity function

The ability to discriminate detail has traditionally been
evaluated by means of alphanumeric optotypes, e.g., the familiar
Snellen test chart or the Landolt C. These stimuli can be used to
measure the ability to resolve high contrast at higher spatial
frequencies. Alternatively, the contrast sensitivity function (CSF)
has been widely utilized in recent years in order to evaluate the
functional range of spatial frequencies. Figure 6 illustrates typical data which indicate the minimum contrast required for threshold
discrimination of sine wave targets of various spatial frequencies.
Application of this more comprehensive measure has uncovered some
unexpected relationships with important implications. Figure 6 presents the CSF for Air Force pilots who have excellent vision as
determined with conventional optotypes. Figure 7 describes the CSF
for groups of healthy observers of different ages. Note that for the
older observers, contrast sensitivity is reduced for the medium and
higher spatial frequencies. In such cases this loss of sensitivity
would have performance implications.
Since medium spatial frequencies are most effective in stimulating
the accommodation reflex (Owens, 1980), such deficits, in the case
of younger observers such as the pilots, might interfere generally
with the ability to accommodate accurately. The loss of medium frequency sensitivity could reduce the distance at which one may appreciate large low contrast objects such as pedestrians. Several
active researchers in this field, particularly Ginsburg and Sekuler,
have advocated that selection and licensing should assess a wider
range of spatial frequencies. Fortunately, the techniques which
they have developed provide a convenient and accurate method for
achieving this noteworthy objective.

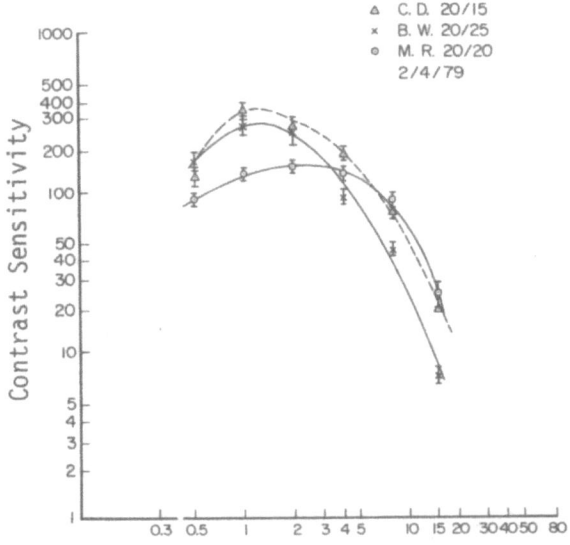

Fig. 6. The contrast sensitivity function of three pilots. Although
   pilots CD and MR have normal Snellen acuity, significant
   differences in their contrast sensitivity are found.
   After Ginsburg, 1981.

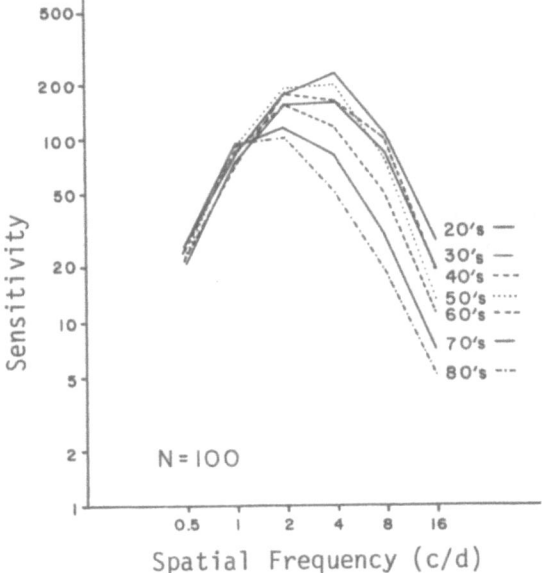

Fig. 7. Contrast sensitivity functions for different age groups.
   After Sekuler and Owsley, 1982.

4.4. Fixational stability

It has been established that the visual measure which correlates most highly with automobile accidents is dynamic visual acuity
(Shinar, 1978). Dynamic acuity is evaluated by determining the
finest spatial detail which can be resolved for a test-object,
moving in the fronto-parallel plane, which the subject is instructed
to track with his eyes. The traditional measure of static acuity is
similar, except that the test-object is stationary. It is interesting
that static acuity does not correlate with either dynamic acuity or
with the frequency of automobile accidents. The high correlation
between dynamic acuity and accident frequency is even more surprising in view of the fact that the task of recognizing fine detail
moving at the velocities typically employed in laboratory tests is
rarely encountered while driving. Rather, the relative motion between the driver and fixated objects on the road typically involves
much slower speeds. This discrepancy as well as the strikingly low
correlation between traditional measures of static acuity and
dynamic acuity (Shinar 1978) suggests that it is not the resolving
power of the eye per se which limits dynamic acuity, but rather
some other visual function related to accident frequency which has
not yet been identified.

In this context, it is possible that dynamic acuity correlates
with accident frequency because it may be an indirect measure of
fixational stability. Errors while attempting to fixate when moving
would result in retinal image smear and eccentric fixation, thereby
degrading dynamic acuity independent of the ability to resolve a
static target. Fixational stability would also be expected a priori
to play a critical function in driving for two additional reasons:
1) The necessity for accurate fixation in maintaining stereoscopic
vision; 2) The contribution of fixational accuracy to the perception
of velocity. In order to elucidate this latter relationship, it is
helpful to review briefly some aspects of the mechanisms underlying
fixational stability.

Fixational stability is subserved by two independent smooth
eye movement systems (Robinson, 1976; Dichgans, 1977). The older
system maintains fixation reflexively via multiple inputs including
vestibular (vestibulo-ocular reflex) and visual (Stiernystagmus or
optokinetic stabilization of gaze). When locomoting or in a moving
vehicle, this system compensates for the effects of head motions by
generating eye movements to maintain retinal image stability. The
other smooth eye movement system (Schaunystagmus or pursuit), which
developed later in evolution, serves primarily to pursue moving
objects. It is characteristic of this system that its activity provides an efference copy (corollary discharge) which serves as the
basis for object motion perception when moving stimuli are pursued
(von Holst and Mittelstaedt, 1950). During normal pursuit of a
moving stimulus the fixated object remains relatively stable on
the retina and the veridical perception of object motion results
from the self-generated efference copy. Under conditions in which

the observer is moving, however, illusory motion perception may occur if the reflex response of the older system is inadequate to maintain fixation. The subsequent fixation errors can be corrected by the pursuit system but this in turn may result in illusory motion perception due to the associated efference copy. This mechanism has been implicated in a number of illusory movement phenomena such as the oculogyral and oculogravic illusions, in which stimulation of the vestibular system results in a tendency to lose fixation which must be opposed by the pursuit system (Whiteside, et al., 1965). In this case the efference copy associated with the pursuit effort required to maintain fixation on a stationary stimulus inappropriately results in motion perception. A similar argument can be made for absolute motion parallax (Post and Leibowitz, 1982) induced motion, and auto-kinesis (Post, et al., 1982).

To the extent that the pursuit effort corresponds to actual object motion, veridical perception would be expected. However, if the effort must also override or compensate for fixational instability, illusory motion, distortions of perceived velocity and distance and possibly spatial location would be anticipated. Such perceptual errors would be expected to degrade driving performance.

Fixational stability is, of course, a fundamental requirement in vehicle guidance. Vehicle motion produces complex displacements of the head and provides both vestibular and visual inputs to the older smooth eye movement system. Errors in fixation, resulting from inadequacies of this system, would interfere with driver performance by degrading resolution and stereopsis. Activation of the pursuit system to compensate for such inadequacy would produce further interference due to perceptual distortions resulting from the associated efference copy. In addition to dynamic acuity, any test which measures fixational instability such as nystagmus induced either by vestibular or optical stimuli, or the stimulus situations described above which cause illusory movement, might be useful in the assessment of driver licensability.

4.5. Visual vestibular interactions

Additional implications within the context of transportation follow from recent demonstrations that object motion perception is affected by concurrent self-motion (Büchele, et al., 1980; Degner and Brandt, 1981). Typically, absolute thresholds of object motion detection are determined for stationary subjects with the head immobile. Under natural conditions, however, humans move freely and experience simultaneous self-motion and object-motion. The thresholds for perception of object motion are raised approximately threefold during sinusoidal head oscillations at 1 Hz about the vertical axis (± 20°) and by a factor of more than six for 2 Hz oscillations respectively (Büchele, et al., 1980). Keeping the head stationary and testing the effects of a purely optokinetically induced self-motion (circular vection, 60°/sec) reveals similar increases in thresholds for object motion detection by a factor of

more than five if apparent self-motion was perpendicular to the
direction of target motion, and almost eighteen for target motion
in the same direction (Degner and Brandt, 1981).

4.6. Aircraft simulation design

Simulators play a major role in the training of pilots by providing an opportunity to practice emergency procedures which would
not be feasible in aircraft. As a result of the increasing complexity of aircraft and rising fuel costs, simulators are expected to
play an even greater role in training and in maintaining proficiency.
The basic problem in simulation design is one of transfer: Does experience in the simulator permit the pilot to perform more effectively in the real aircraft? Since no simulator can reproduce all of
the stimuli encountered in a real aircraft, the problem reduces to
predicting the effect on transfer of experience in a sensory environment with limited fidelity. This decision must, in turn, be based on
appreciation of the role of basic mechanisms of spatial orientation.
For example, since spatial orientation depends on the interaction
among sensory subsystems, what is the effect on pilot performance of
eliminating, or partially eliminating, one of these inputs during
training? This is illustrated by the question of whether a simulator
should include a moving base in addition to visual stimuli. Motion
imparted by such a base cannot correspond completely to the accelerations in an aircraft so that only some of the vestibular stimulations are feasible. Because of the tremendous costs involved in
building moving base simulators, consideration has been given to
eliminating the moving base. We know that selective elimination of
sensory inputs can lead to disorientation but this may, with practice, be overcome by adaptation. If, however, a pilot who has adapted to exclusively visual inputs is subsequently exposed to the
normal pattern of interaction in a real airplane, a mismatch will
result with possible disorientation and motion sickness. Whatever
the solution to this complex problem, it is clear that it will
depend on an understanding of basic spatial orientation mechanisms.

Vehicle guidance involves both the focal and ambient modes of
processing which have different psychophysical properties. Given the
fact that ambient vision has only minimal optical fidelity requirements, it may be feasible to reduce the resolution of the computer
generated displays in simulators for those visual stimuli which
provide orientation cues. This is clearly not the case for focal
processing. However, considerable savings would be possible if the
resolution requirements for orientation stimuli could be reduced.

The peripheral visual fields play a major role in spatial
orientation. In a simulator, the question of how much of the peripheral visual field should be stimulated is important both with
respect to transfer of training and economic considerations. The
smaller the visual field the less the cost, but this "tradeoff"
question can be addressed only in the context of the role of stimulation of the periphery in training and its relationship to actual

performance. Considerations of this problem are currently limited by a lack of knowledge of the role of specific areas of the visual fields in orientation.

A number of studies suggest that during psychological or physiological stress, the functional visual fields are narrowed (Leibowitz, 1973) but the implications of this literature are not clear. We have suggested the possibility that under some kinds of stress, narrowing may be limited to focal processing while ambient functions remains intact (Leibowitz, et al., 1980). This problem is important in simulation because stress is commonly encountered in military aviation and in space flight. If, for example, both ambient and focal modes of processing were lost from the far periphery during stress, it may be undesirable to stimulate these regions of the retina during simulation training. Alternatively, if the gain of visual stimulation is altered during stress it could result in a mismatch-related disorientation. In this case, it might be desirable to attempt to train under mismatch conditions. Unfortunately, present methods of evaluating the extent of the peripheral visual fields are restricted to measures indicating focal functions, measures which are not sensitive to ambient vision. For a number of reasons, in particular the possible role of stress, ambient as well as focal vision should be evaluated in visual field testing. This is important since stress can have many origins. It is possible that "psychological" stressors such as fear may result in a narrowing of focal vision only. There are no direct data on this problem but its feasibility is suggested by observations of psychiatric patients among whom orientation appears normal in spite of severe narrowing (Jolly, 1892; Janet, 1907). On the other hand, for stressors such as hypoxia and gravity forces which have been reported to produce narrowing, the mechanisms may be different, e.g. restriction of peripheral retinal blood flow, so that narrowing would be expected to involve both ambient and focal vision.

Clearly, the contribution of different areas of the visual fields to spatial orientation and the effects of stressors are important in simulator design. In order to maximize the transfer of training from simulators to real aircraft and to minimize the costs of construction, additional understanding of dynamic spatial orientation mechanisms is both wanted and wanting.

4.7. Motor skills

Although this chapter is concerned mainly with the analysis of visual perception as a factor in vehicle guidance, it should be mentioned here that optimal performance on such tasks as car driving also involves other capacities. One of the most important among these is the capacity to correctly perform certain motor skills.

The neural calibration of motor control and vehicle guidance mechanisms results from repeated specific exposures. The learning process of riding a bike or driving a car is characterized by the establishment or calibration of new preprogrammed neural patterns

specific to the static and dynamic properties of the vehicle. The
necessary angle of rotation of the steering wheel must be adjusted
to the desired directional corrections of the moving vehicle. The
differential muscle force required either for a slow visually controlled deceleration or a rapid emergency braking must be adjusted
to the resistance force of the foot pedal. The availability of a
neural store providing motor program patterns for several reflex-
like reactions undoubtedly facilitates vehicle guidance. It may,
however, cause difficulty if a driver changes vehicles. Then, in an
emergency situation, automatic steering and braking reactions will
be released due to prior experience but they will be inadequate to
control the new vehicle because of the different dynamic properties.
The driver moving from one car to another does not realize this
potential risk because his self-confidence is supported by his
ability to control the car in normal traffic by slow, visually-
guided maneuvers different from the preprogrammed reactions. This
neurophysiological mechanism is the basis for the well known difficulties in balance that arise if somebody able to ride a bicycle
tries to ride a heavy motorcycle. Such inappropriate vehicle guidance
could be reduced: 1. by standarization of the dynamic properties of
vehicles within one license group; 2. by training of vehicle control
under emergency situations both in driving school and later when
first entering a new vehicle; 3. by instructing drivers about the
selective nature of sensory-motor calibration and vehicle guidance.

Conclusion

     To reiterate the main theme of this chapter, the material pre-
sented above argues against maintaining a distinction between "basic"
and "applied" research. Rather, they play a mutually supportive
role since progress in the solution of societal problems frequently
depends on our knowledge of fundamentals. Similarly, societal
problems often highlight weaknesses in our appreciation of basic
mechanisms and thus serve a valuable function in directing research
efforts.

REFERENCES

Allen, M.J., Vision and Highway Safety. Philadelphia: Chilton,
     1970.
Baldwin, W.R. and Stover, W.B., Observation of laser standing wave
     patterns to determine refractive status. American Journal of
     Optometry, 1968, 45, 143-150.
Baumgarten, von, R.J. and Thümler, R.A., A model of vestibular
     function in altered gravitational states. In: R. Holmquist,
     (Ed.): (Cospar) Life Sciences and Space Research. Oxford,
     England, Pergamon Press, 1970, 161-170.
Benson, A.J. Possible mechanisms of motion and space sickness.
     Proceedings of the European Symposium on Life Sciences Research
     in Space, European Space Agency, 1977 Sp-130, 101-108.

Benson, A.J., In: Aviation Medicine, Physiology and Human Factors,
     G. Dhenin, G.E. Sharp and J. Ernsting (Eds.). London: Tri-Med
     Books, I., 1978, Spatial Disorientation, Chapters 20 and 21.
Brandt, Th., Optisch-vestibuläre Bewegungskrankheit, Höhenschwindel
     und klinische Schwindelformen. Fortschritte der Medizin, 1976,
     94, (20-21), 1177-1182.
Brandt, Th. Medikamentöse und physikalische Therapie des Schwindels
     und der Ataxie. Fortschritte der Neurologie und Psychiatrie,
     1981, 49(3), 81-100.
Brandt, Th., Wist, E.R. and Dichgans, J., Foreground and background
     in dynamic spatial orientation. Perception and Psychophysics,
     1975, 17, 497-503.
Brandt, Th., Wenzel, D. and Dichgans, J., Die Entwicklung der
     visuellen Stabilisation des aufrechten Standes beim Kind: Ein
     Reifezeichen in der Kinderneurologie, Archiv Psychiatrie und
     Nervenkrankheiten, 1976, 223, 1-13.
Brandt, Th. and Daroff, R.B., The multisensory physiological and
     pathological vertigo syndromes. Annals of Neurology, 1980,
     7(3), 195-203.
Büchele, W., Degner, D. and Brandt, Th., Thresholds for object-
     motion perception raised by concurrent head movements.
     Pflügers Archiv, Supplement, 1980, R33, 384.
Chinn, H.I. and Smith, P.K., Motion Sickness. Pharmacological Review,
     1955, 7, 33-83.
Degner, D. and Brandt, Th., Interaction between self- and object-
     motion perception. Pflügers Archiv, 1981, in press.
Dichgans, J. Optokinetic nystagmus as dependent on the retinal peri-
     phery via the vestibular nucleus. In: Control of Gaze by Brain
     Stem Neurons, R. Baker and A. Berthoz (Eds.). North Holland/
     Elsevier, Amsterdam, 1977, pp. 261-267.
Dichgans, J. and Brandt, Th., Optokinetic motion sickness and
     pseudo-coriolis effects induced by moving visual stimuli.
     Acta Otolaryngologica, 1973, 76, 339-348.
Dichgans, J., Schmidt, C.L. and Graf, W. Visual input improves the
     speedometer function of the vestibular nuclei in the goldfish,
     Experimental Brain Research, 1973, 18, 319-322.
Dichgans, J. and Brandt, Th., Visual-vestibular interaction: Effects
     on self-motion and postural control. In: Handbook of Sensory
     Physiology, (Vol. VIII), by R. Held, H.W. Leibowitz and
     H.L. Teuber (Eds.), Heidelberg: Springer, 1978.
Ginsburg, A.P., Spatial filtering and vision: Implications for normal
     and abnormal vision. In: Clinical Applications of Visual Psycho-
     physics, L. Proenza, J. Enoch and A. Jampolsky (Eds.), Cambridge
     University Press, 1981.
Gonshor, A. and Melville-Jones, G., Short term changes in the human
     vestibulo-ocular reflex. Journal of Physiology, London, 1976,
     226, 361-379.
Gramberg-Danielsen, B. Sehen und Verkehr. Berlin: Springer, 1967.
Held, R., Dissociation of visual functions by deprivation and re-
     arrangement. Psychologische Forschung, 1968, 31, 338-348.

Held, R., Two modes of processing spatially distributed visual
    stimulation. In: F.O. Schmidt (Ed.), The Neurosciences:
    Second Study Program. New York, Rockefeller University Press,
    1970.

Hennessy, R.T. and Leibowitz, H.W., Subjective measurement of accom-
    modation with laser light. Journal of the Optical Society of
    America, 1970, 60, 1700-1701.

Holst, E.V. and Mittelstaedt, H., Das Reafferenzprinzip, Wechsel-
    wirkungen zwischen Zentralnervensystem und Peripherie. Natur-
    wissenschaften, 1950, 37, 464-476.

Janet, P. The major symptoms of hysteria. London, Macmillan, 1907.

Johnson, C.A., Effects of luminance and stimulus distance on
    accommodation and visual resolution. Journal of the Optical
    Society of America, 1976, 66, 138-142.

Jolly, F., Ueber Hysterie bei Kindern. Berliner Klinische Wochen-
    schrift, 1892, 29, 841-845.

Kaufman, L., Sight and mind. New York, Oxford, 1974.

Knoll, H.A., Measuring ametropia with a gas laser: A preliminary
    report. American Journal of Optometry, 1966, 43, 415-418.

Leibowitz, H.W., Detection of peripheral stimuli under psychological
    and physiological stress. In: Visual Search. Washington, D.C.,
    Committee on Vision, National Research Council-National Academy
    of Sciences, 1973.

Leibowitz, H.W., Myers, N.A. and Grant, D.A., Radial localization of
    a single stimulus as a function of luminance and duration of
    exposure. Journal of the Optical Society of America, 1955,
    45(2), 76-78.

Leibowitz, H.W. and Owens, D.A., Anomalous myopias and the inter-
    mediate dark focus of accommodation. Science, 1975, 189,
    646-648.

Leibowitz, H.W. and Owens, D.A., Nighttime accidents and selective
    visual degradation. Science, 1977, 197, 4302, 422-423.

Leibowitz, H.W. and Owens, D.A., New evidence for the intermediate
    position of relaxed accommodation. Documenta Ophthalmologica,
    1978, 46(1), 133-147.

Leibowitz, H.W., Shupert-Rodemer, C. and Dichgans, J., The indepen-
    dence of dynamic spatial orientation from luminance and re-
    fractive error. Perception and Psychophysics, 1979, 25(7),
    75-79.

Leibowitz, H.W., Post, R.B., Shupert-Rodemer, C., Wadlington, W.L.
    and Lundy, R.M., Roll vection analysis of suggestion induced
    visual field narrowing. Perception and Psychophysics, 1980,
    28(2), 173-176.

Leibowitz, H.W. and Dichgans, J., The ambient visual system and
    spatial orientation. Proceedings of the AGARD conference on
    spatial disorientation in flight, Bodø, Norway, May, 1980.

Leibowitz, H.W. and Post, R.B., The two modes of processing concept
    and some implications. Organization and Representation in
    Perception, J. Beck (Ed.), Erlbaum, 1982, in press.

Levene, J.R., Nevil Maskelyne, F.R.S. and the discovery of night myopia. Roy. Soc. Lond. Notes and Reports, 1965, 20, 100–108.

Luria, S.M., Target size and correction for empty field myopia.
    J. of the Optical Society of America, 1980, 70(9), 1153–1154.

Malcolm, R., Money, K.E. and Anderson, P., Peripheral vision artifi-
    cial display. AGARD Conference Proceedings No. 145 on Vibration
    and Combined Stress in Advanced Systems, 1975, B20-1-B20-3.

Money, K.E., Malcolm, R.E. and Anderson, P.J., The Malcolm horizon,
    AGARD conference Proceedings No. 201 on visual presentation of
    cockpit information including special devices use for particular
    conditions of flying, 1976, A41–A43.

Money, K.E., Motion sickness. Physiological Reviews, 1970, 50, 1–39.

Owens, D.A., The Mandlebaum Effect: An accommodative response bias
    toward intermediate distance. Journal of the Optical Society of
    America, 1979, 69, 646–652.

Owens, D.A., A comparison of contrast sensitivity and accommodative
    responsiveness for sinusoidal gratings, Vision Research, 1980,
    20, 159–167.

Owens, D.A. and Leibowitz, H.W., Night myopia: Cause and a possible
    basis for amelioration. American Journal of Optometry and
    Physiological Optics, 1976, 53, 709–717.

Perenin, M.T. and Jeannerod, M., Residual vision in cortically blind
    hemifields. Neuropsychologia, 1975, 13, 1–7.

Perenin, M.T. and Jeannerod, M., Visual function within the hemiano-
    pic field following early cerebral hemidecortication in man-I.
    Spatial localization. Neuropsychologia, 1978, 16, 1–13.

Pöppel, E., Held, R. and Frost, D., Residual vision after brain
    wounds involving the central visual pathways in man. Nature,
    London, 1973, 243, 295–296.

Post, R.B., Stimulus control of circular vection and optokinetic
    afternystagmus. Doctoral dissertation. Pennsylvania State
    University, 1982.

Post, R.B., Owens, R.L., Owens, D.A. and Leibowitz, H.W., Correction
    of empty field myopia on the basis of the dark-focus of accom-
    modation. Journal of the Optical Society of America, 1979,
    69(1), 89–92.

Post, R.B. and Leibowitz, H.W., The independence of radial localiza-
    tion from refractive error. Journal of the Optical Society of
    America, 1980, 70(11), 1377–1379.

Post, R.B. and Leibowitz, H.W., The effect of convergence on the
    vestibulo-ocular reflex and implications for perceived movement,
    Vision Research, in press, 1982.

Post, R.B., Shupert, C.L. and Leibowitz, H.W., Autokinesis and
    peripheral stimuli: Implications for fixational stability,
    Perception, in press, 1982.

Probst, Th., Krafczyk, S., Büchele, W. and Brandt, Th. Visuelle
    Prävention der Bewegungskrankheit im Auto. Fortschritte der
    Medizin, 1981, in press.

Reason, J.R. and Brand, J.J., Motion Sickness. London/New York:
    Academic Press, 1975.

Robinson, D., The physiology of pursuit eye movements. In: Eye Move-

ments and Psychological Processes, by R.A. Monty and J.W.
    Senders (Eds.). Erlbaum, Hillsdale, New Jersey, 1976.
Schneider, G.E., Contrasting visuomotor functions of tectum and
    cortex in the golden hamster. Psychologische Forschung, 1967,
    31, 52-62.
Schober, H.A.W., Ueber die Akkommodationsruhelage. Optik, 1954,
    11, 282-290.
Sekuler, R. and Owsley, C. Spatial Vision in Older Humans. In:
    Aging in Human Visual Functions, by R. Sekuler, D. Kline and
    K. Dismukes (Eds.). Alan R. Liss, Inc., New York, 1982 (in
    press).
Shinar, D., Psychology on the Road. New York: Wiley, 1978.
Slovic, P., Fischhoff, B. and Lichtenstein., Accident probabilities
    and seat belt usage: A psychological perspective. Accident
    Analysis and Prevention, 1978, 10, 2181-285.
Waespe, W. and Henn, V., Behaviour of secondary vestibular units
    during optokinetic nystagmus and after-nystagmus in alert
    monkeys. Pflügers Archiv, Supplement R, 1976, 362, 197.
Weiskrantz, L., Warrington, E.K., Sanders M.D. and Marshall, J.,
    Visual capacity in the hemianopic field following a restricted
    occipital ablation. Brain, 1974, 97, 709-728.
Whiteside, T.C.D., Accommodation of the human eye in a bright and
    empty field. Journal of Physiology (London), 1952, 118, 65.
Whiteside, T.C.D., Graybiel, A. and Niven, J.I., Visual illusions
    of movement, Brain, 1965, 88, 193-210.

INDEX